Nanomaterials for Energy Applications

Topical Conference at the 2008 AIChE Annual Meeting

Philadelphia, Pennsylvania
16 - 21 November 2008

ISBN: 978-1-61567-209-7

Printed from e-media with permission by:

Curran Associates, Inc.
57 Morehouse Lane
Red Hook, NY 12571

Some format issues inherent in the e-media version may also appear in this print version.

Printed by Curran Associates, Inc. (2009)

For permission requests, please contact AIChE
at the address below.

AIChE
3 Park Avenue
New York, NY 10016-5991

Phone: (203) 702-7660
Fax: (203) 775-5177

www.aiche.org

Additional copies of this publication are available from:

Curran Associates, Inc.
57 Morehouse Lane
Red Hook, NY 12571 USA
Phone: 845-758-0400
Fax: 845-758-2634
Email: curran@proceedings.com
Web: www.proceedings.com

TABLE OF CONTENTS

45a Morphological Dependence of ZnO on Methanol Steam Reforming Activity in Pd/Zno Catalysts

Patrick D. Burton[1], Travis Conant[1], Chuan Zhang[1], Abhaya K. Datye[1], Bernadette A. Hernandez-Sanchez[2], and Timothy J. Boyle[2]. (1) Chemical and Nuclear Engineering, University of New Mexico, 1 The University of New Mexico, MSC01 1120, Albuquerque, NM 87131, (2) Inorganic Chemistry and Nanomaterials - 1846, Sandia National Laboratories, Advanced Materials Laboratory - MS 1349, 1001 University Blvd. S.E., Albuquerque, NM 87106

$Pd/ZnO/Al_2O_3$ catalysts have been shown to be highly active for methanol steam reforming and selective towards production of CO_2 and H_2. Previous work has shown that PdZnO catalysts prepared from an acidic precursor can etch the support, destroying the faceted morphology of the ZnO support, and leading to a decrease in activity. Preserving the ZnO facets through use of an organic precursor leads to improved catalytic activity. It is not clear wether ZnO participates in this reaction through a bifunctional pathway. This work addresses the role of ZnO by systematically investigating the catalytic activity of samples prepared from faceted ZnO of differing morphology (nanorods and platelets), etched and reprcipitated ZnO, and $ZnAl_2O_4$. Pd deposition on each support was carried out by using incipient wetness with a $Pd(OAc)_2$ precursor. An alternative catalyst preparation using preformed nanoparticles was also investigated. We used CO oxidation activity to quantitatively determine the number of active Pd sites, and methanol steam reforming (after forming PdZn particles *in-situ*). These results show that the ZnO surface morphology plays a significant role in afecting the reactivity of Pd based steam reforming catalysts.

This work has been supported by the United States Department of Energy, Office of Basic Energy Sciences under contract number DE-FG02-05ER15712 (University of New Mexico) and DE-AC0494AL85000 (Sandia National Laboratories). Sandia is a multiprogram laboratory operated by Sandia Corporation, a Lockheed Martin Company for the United States Department of Energy.

45b Tailoring the Pore Morphology of Pure and Mixed La- and Ce-Oxides

Yanan Wang[1], Shuang Liang[1], and Götz Veser[2]. (1) Chemical and Petroleum Engineering Department, University of Pittsburgh, 3700 O Hara Street, Pittsburgh, PA 15261, (2) Department of Chemical Engineering, University of Pittsburgh, Pittsburgh, PA 15261

Advances in materials chemistry over the past decade have opened new possibilities for tailoring the properties of technical catalysts from the nanoscale up. Nanoporous materials, with well-defined pore morphologies, i.e. controlled pore diameter and structure, give rise to large and highly accessible internal surface areas, as well as unique properties such as molecular sieving capabilities and/or the capability to integrate multiple functionalities into a single material.

We are reporting on the synthesis, characterization, and testing of nanostructured CeO2- and La2O3-based catalysts for the Water-Gas Shift (WGS) reaction. The work builds onto our previous work in which we demonstrated the unusual stability of noble metal/alumina nanocomposite materials in high-temperature catalysis, such as catalytic partial oxidation of methane to synthesis gas. Target of the present work is the synthesis of similar nanocomposites by embedding metal nanoparticles into a nanostructured CeO2 and/or La2O3 matrix. Both oxides have previously been shown to be effective (and active) support materials for noble metal-based catalysts in WGS. However, previous work has mostly focused on the effect of "nano-sizing" the active metal component in these catalysts, although recent evidence indicates that the structure of the oxide support is much more important than given credit to-date.

Our synthesis is based on a reverse microemulsion-templated sol-gel synthesis, in which the embedded metal nanoparticles are synthesized simultaneously with the ceramic matrix. This templating route allows for a flexible tailoring of the catalyst morphology by manipulation of a variety of synthesis parameters. In this way we have successfully synthesized a range of pure and mixed CeO2- and La2O3-based nanostructured ceramic supports with high surface area and good thermal stability over a broad temperature range of relevance for WGS. The resulting materials were characterized via SEM, TEM, XRD, BET, and chemisorption. We found that the pore size and surface area of pure CeO2 or La2O3 materials can be tailored via synthesis parameters such as the water:alkoxide ratio. The synthesis of mixed CeO3-La2O3 provides not only higher surface area than either of the pure oxides over the relevant temperature range, but the Ce:La ratio also has a profound impact on the pore morphology of these resulting materials.

Finally, Au, Ni, and Pt nanoparticles were successfully embedded into these nanostructured ceramic supports with different pore morphologies. We are currently evaluating the activity and stability of these nanostructured catalysts in WGS. First results indicate excellent stability of these nanocomposite catalysts. Synthesis, characterization, and reactive tests will be discussed in detail in the presentation.

45c α-Fe2O3 Nanowire Array Based Electrodes for Self-Driven Photoelectrochemical Cells

Boris D. Chernomordik, Chemical Engineering, University of Louisville, 106 Ernst Hall, Department of Chemical Engineering,, University of Louisville, Louisville, KY 40292, Uros Cvelbar, Plasma Laboratory F4, Jozef Stefan Institute, Jamova 39, Ljubljana, SI-1000, Slovenia, Vidhya Chakrapani, Chemical Engineering, University of Lousiville, Room 311, Ernst Hall, Univeristy of Louisville, Louisville, KY 40292, Miran Mozetic, Department for Surface Engineering, Jozef Stefan Institute, Jamova 39, Ljubljana, Slovenia, and Mahendra K. Sunkara, Department of Chemical Engineering, University of Louisville, 106 Ernst Hall, Department of Chemical Engineering,, University of Louisville, Louisville, KY 40292.

Iron oxide, Fe2O3, is a promising material for water splitting reaction using solar energy.[1-2] Even the recent efforts, however, using Fe2O3 thin film materials reported low efficiencies due to poor carrier transport within these films.[3-4] In this regard, we have been working on a-Fe2O3 nanowire arrays as possible thin film materials systems for improved carrier mobility during photoelectrochemical water splitting reaction.

Recently, we developed a novel method to synthesize arrays of α-Fe2O3 nanowires, which are single crystal and have highly ordered oxygen vacancy plane.[5-6] The nanowire arrays are directly synthesized on metallic substrates using low-temperature RF plasma oxidation without the use of any foreign metal catalysts. The nanowire growth is spontaneous and the process lasts only seconds. This method could be scaled up for manufacturing of large area nanowire arrays on metallic foils. As one-dimensional nanostructures, these nanowires offer many other benefits to PEC electrolysis, such as high surface area, reduced charge carrier diffusion distance, and a preferential direction for charge diffusion. In this presentation, we will provide both synthesis and their photoelectrochemical properties of a-Fe2O3 nanowire arrays grown on Fe foils.

References

1. K.L. Hardee, A.J. Bard, Semiconducting electrodes. 5. Application of Chemically Vapor Deposited Iron Oxide Films to Photosensitized Electrolysis", J. of the Electrochemical Society 1976, 123, (7), 1024-1026.

2. J.H. Kennedy, M. Anderman, Photoelectrolysis of Water at alpha- Fe2O3 electrodes in Acidic Solution", J. of the Electrochemical Society 1983, 130, (4),

3. T. Lindgren, H.L. Wang, N. Beermann, L. Vayssieres, A. Hagfeldt, S.E. Lindquist, Aqueous photoelectrochemistry of hematite nanorod array, "Solar Energy Materials and Solar Cells", 2002, 71, (2), 231-243.

4. W.B. Ingler, S.U.M. Khan, "A self-driven p/n-Fe2O3 tandem photoelectrochemical cell for water splitting", Electrochemical and Solid State Letters 2006, 9, (4), G144-G146.

5. U. Cvelbar, Z. Chen, M.K. Sunkara and M. Mozetic, "Spontaneous growth of superstructure α-Fe2O3 nanowire and nanobelt arrays in reactive oxygen plasma", Revision Submitted, Small (2008).

6. Z. Chen, U. Cvelbar, M. Mozetic, J. He and M.K. Sunkara, "Long-Range Ordering of Oxygen-Vacancy Planes in r-Fe2O3 Nanowires and Nanobelts", Chemistry of Materials, 20 (9), 3224 - 3228 (2008).

45d TiO2 Nanotubes Supported Gold Photoanodes for Photoelectrochemical Hydrogen Production

Peter H. Aurora[1], Levi Thompson[2], and Chang Hwan Kim[2]. (1) Mechanical Engineering, University of Michigan, 2300 Hayward Street, 3216 H.H. Dow Building, Ann Arbor, MI 48109, (2) Chemical Engineering, University of Michigan, 3074 H.H. Dow, 2300 Hayward, Ann Arbor, MI 48105

Photoelectrochemical cells (PEC) combine a photovoltaic device for light harvesting and an electrolyzer for water splitting into a one single device. In its simplest form, a PEC cell is composed of a photoanode for water oxidation where oxygen is evolved, a cathode where hydrogen is evolved, and an electrolyte. A major obstacle for the wide-spread use of PEC cells is the low photoanode efficiency. By some accounts, the rate of water oxidation has to be increased by more than an order of magnitude to keep pace with the production of electrons and holes. In addition, recombination losses can limit performance. We explored two strategies for improving performance of the photoanode: producing the photocatalyst in the form of nanotubes to improve collection efficiencies and incorporating nanocrystalline gold onto the nanotubes to improve activity.

Highly ordered TiO2 nanotube (TiNT) arrays were fabricated using an anodization process. Gold nanoparticles were supported onto the TiNTs using a modified deposition precipitation (DP) method. The pH, aging time and Au precursor concentration were manipulated to increase the Au dispersion and loading on the nanotubes. Cyclic voltammetry was used to measure the total and Au electrochemical surface areas. The performance of photoanodes was evaluated in a 3-electrode cell with a 1.0 M KOH electrolyte using a solar simulator (1.5 AM) and potentiostat. Scanning electron microscopy (SEM), transmission electron microscopy (TEM), and X-ray diffraction were used to characterize key properties of the materials, and optical absorption measurements were used to determine their bandgaps.

Longer nanotubes provided higher photocurrents compared to the TiO2 powders and shorter nanotubes. This was attributed to enhanced capture of the light and better separation of the charge carriers. The deposition of Au nanoparticles resulted in a slight reduction of the bandgap, which we attributed to the existence of impurity levels between the band edges of the oxide. Introduction of the gold nanoparticles resulted in a significant improvement in the electrocatalytic properties. In addition the intrinsic activity increased as the Au particle size decreased, in a manner similar to that observed for CO oxidation. We observed a three-fold increase in activity on reducing the average Au particle size from 28 to 2.9 nm. These and other results will be presented in this paper.

45e Band Gap Modification of Oxide Nanowires for Solar Hydrogen Production

Vidhya Chakrapani, Chemical Engineering, University of Lousiville, Room 311, Ernst Hall, Univeristy of Louisville, Louisville, KY 40292, Jyothish Thangala, Chemical engineering, University of Louisville, 106, Ernst Hall, Louisville, KY 40292, Zhiqiang Chen, Institute for Advanced Materials and Renewable Energy, University of Louisville, Louisville, KY 40292, and Mahendra K. Sunkara, Department of Chemical Engineering, University of Louisville, 106 Ernst Hall, Department of Chemical Engineering,, University of Louisville, Louisville, KY 40292.

The photocatalytic production of hydrogen via water splitting is a subject of immense interest due to its possible applications for converting and storing the abundant sunlight energy as chemical energy. However, with the currently used materials, efficiency of this process has been less than 1%. Improvement in the device performance is limited by either chemical instability or poor light absorption properties of these materials. Transition metal oxides are chemically stable during gas evolution, but have high band gap which limits their use in the UV part of the solar spectrum. In this work, we performed post synthesis nitridation of oxide nanowire arrays to form a completely new nitrided phase which has a lower band gap. The photoelectrochemical properties of these nanowire electrodes were characterized by UV-Vis, impedance, and photocurrent spectroscopy. The results are discussed in terms of their water splitting efficiency.

102a Roles of Catalysts on Hydrogen Desorption of Mixtures of $LiBH_4$ and MgH_2 and Its Reversibility

Pattarapom Sridechprasat[1], Yindee Suttisawat[1], Pramoch Rangsunvigit[1], Boonyarach Kitiyanan[1], and Santi Kulprathipanja[2]. (1) Petrochemical Technology, The Petroleum and Petrochemical College, Chulalongkorn University, Bangkok, 10330, Thailand, (2) UOP LLC, 50 East Algonquin Road, Des Plaines, IL 60017-5016

Effects of catalysts (Ti, TiO_2, and $TiCl_3$) on hydrogen desorption of a 2:1 molar ratio of LiBH4:MgH2 and its reversibility were investigated. 3 mol% of a Ti-compound were added as a catalyst in the hydride mixture using mechanical ball-milling for one hour as a mixing means. Hydrogen desorption was carried out under atmospheric pressure with the temperature range from 25 to 500°C and a heating rate of 2°C min^{-1} while hydrogen re-absorption took place at 5 MPa hydrogen pressure and 350°C. The hydrogen desorption results of the uncatalyzed ($LiBH_4$-MgH_2) and catalyzed mixtures occurred in two steps. It was found that the uncatalyzed and catalyzed mixtures started to release hydrogen at the same temperature of 150°C for the first step and 400°C for the second step. In the presence of a catalyst, the mixtures released hydrogen at a higher amount (6.5 wt%) than that of $LiBH_4$-MgH_2 (3 wt%) in the first step. When the temperature reached 500°C, the hydrogen desorption of both mixtures were approximately the same, 7 wt%. To understand roles of the metal hydrides and the catalysts in each step of the hydrogen desorption, all mixtures were characterized by XRD. The results identified that the hydrogen desorption from the first step was contributed by a reaction between $LiBH_4$ and MgH_2 while that from the second step was from a reaction between $LiBH_4$ and Mg. Furthermore, the catalysts affected the decomposition of MgH_2 in a greater extent than $LiBH_4$. Both $LiBH_4$-MgH_2 and that with any tested catalyst possessed better reversibility than each hydride alone. In addition, the hydrogen desorption and re-absorption using difference types of the catalysts will also be discussed.

102b Synthesize of Nanostructured Lithium Imide and Its Application for Hydrogen Storage

Jizhong Luo[1], Ziyi Zhong[2], Y.S. Loo[2], and L.W. Chen[2]. (1) Institute of Chemical & Engineering Sciences, 1 PESEK ROAD JURONG ISLAND SINGAPORE 627833, Singapore, 627833, Singapore, (2) Applied Catalysis, Institute of Chemical & Engineering Sciences, 1 PESEK ROAD JURONG ISLAND SINGAPORE 627833, Singapore, 627833, Singapore

Nanostructured lithium imide (Li2NH) was synthesized by solid ion exchange reaction of Li3N and LiNH2. The average crystallites diameter is around 12 nm. Hydrogen adsorption study indicated that the enthalpy change for the hydrogenation over this material is ~ 35 KJ/mol, which is lower than the reported ones (-60-65 KJ/mol), also lower than theoretical prediction value (~ 45 KJ/mol). This enthalpy change as well as 6.5 wt% of reversible hydrogen capacity suggests that this lithium imide is a good candidate for on-board hydrogen storage. Further modifying with Mg, Ti, etc over lithium imide did not increase the corresponding hydrogen adsorption activity while the cycle stability was markedly enhanced.

Dehydrogenation Thermodynamics of Hydrogen Storage Material Manganese Borohydride: First-Principles Study

Pabitra Choudhury[1, 2], Venkat R. Bhethanabotla[1, 2] and Elias Stefanakos[2]
[1]Sensors Research Laboratory, Department of Chemical and Biomedical Engineering
[2]Clean Energy Research Center, Department of Electrical Engineering
University of South Florida, 4202 E. Fowler Ave., ENB 118, Tampa, FL 33620

Abstract

Ab initio calculations using DFT were performed to establish a stable crystal structure of $Mn(BH_4)_2$. The crystal structure of $Mn(BH_4)_2$ is found to be orthorhombic with space group *Fddd* (No. 70). The electronic density of states shows that manganese borohydride is of metallic nature without a band gap. Our calculations showed that the interaction between Mn atoms and BH_4 complexes has an ionic character while the internal bonding of BH_4 is essentially covalent. The reaction enthalpy was also found for the reaction $Mn(BH_4)_2 = Mn + 2B + 4H_2(g)$ to be 135.026 kJ/f.u. at 0 K including zero point energy (ZPE) corrections.

Introduction

Light weight, low cost, highly reversible hydrogen storage systems are essential for low temperature PEM fuel cell powered vehicles[1]. Complex hydrides of alkali and alkaline earth elements are promising candidates for hydrogen storage, but typically have heats of reaction that are too high to be of use for fuel cell vehicles. So, transition metal borohydride complexes as hydrogen storage materials have recently attracted great interest. $Mn(BH_4)_2$, among all other transition complex borohydrides, is stable at room tempearature[2], and is considered to be a potential candidate for on-board applications as it has high theoretical hydrogen storage capacity of 10 wt% and also a reasonably low decomposition temperature compared to complex alkali and alkaline earth borohydrides. To be a viable candidate, a hydride must have a reaction free energy that lies in a range of values to allow reversible H_2 storage at practical temperatures and pressures. Thermodynamic data such as standard enthalpy of formation and Gibbs energy of dehydrogenation reaction required to assess the usability of the candidate material for practical applications are not available.

A combined theoretical and experimental approach can better lead us to design a suitable complex material for hydrogen storage. Here, we explore the complex hydrides using density functional theory (DFT) to provide information at the electronic level that is hard to obtain by experimental methods. Such calculations can also unravel the transition states of the molecules undergoing chemical transformation at complex solid surfaces. Phonon spectrum helps us to calculate the finite temperature thermodynamics and such calculations are performed in this work.

Computational details and methodology

First-principle calculations were performed in this work on $Mn(BH_4)_2$ using DFT within the generalized gradient approximation (GGA) and projector augmented wave (PAW) method and a plane wave basis set to calculate the total energies, as implemented in the Vienna *Ab initio* Simulation Package (VASP).[3-5] We started our calculations from known similar complex formula units $X(YZ_4)_2$[6] for $Mn(BH_4)_2$ to find a stable crystal structure. Based on the optimized crystal structure from DFT, lattice dynamics based on the harmonic approximation using the direct force constant method in the PHONON[7] code was used to calculate the phonon density of states (DOS) to confirm the finite temperature crystal stability and then calculated finite temperature thermodynamics.

The standard enthalpy of formation is an important predictor of decomposition temperature for the decomposition reaction with its associated standard thermodynamic parameters. For a complex hydride, when the material is being heated, the effect of standard entropy starts dominating the standard enthalpy and at the decomposition temperature and constant pressure the standard Gibbs energy is zero. As during heating the entropy change of solid materials is very small compared to the gas, we can consider that the entropy change during the decomposition reaction is primarily due to H_2 evolution. We have calculated the enthalpy for the formation equation of $Mn(BH_4)_2$ from its elemental components as given below:

$$Mn + 2B + 4H_2(g) = Mn(BH_4)_2$$

The calculated results indicated that those for space group *Fddd* gives the stable structure. We have also calculated the electronic total density of states (DOS) and Electron Localization Function (ELF) of $Mn(BH_4)_2$ to gain insights into the nature of the material and bonding between the elements.

Results/Discussion

The above methods were used to establish the stable crystal structure with appropriate lattice parameters & atomic positions and the electronic structure of the high theoretical hydrogen storage capacity material $Mn(BH_4)_2$. The minimum total energy for $Mn(BH_4)_2$ was found to be -49.510 eV/mol at a unit cell volume of 816.10 Å^3. The final optimized crystal structure with atomic positions is shown in Table 1.

Table 1: Optimized crystal structure of $Mn(BH_4)_2$. The space group is *Fddd* (# 70) with lattice parameters a = 9.1562 Å, b = 12.8508 Å and c = 6.9358. All the atomic positions are given by Wyckoff letter and the number of formula units in the unit cell *Z*=8.

Element	Wyckoff letter	X	Y	Z
Mn	*8a*	0.0000	0.0000	0.0000
B	16f	0.0000	0.1917	0.0000
H1	32*h*	0.8959	0.7493	0.4659
H2	32*h*	0.9648	0.6399	0.6414

The crystal structure of $Mn(BH_4)_2$ is consist of metal cation Mn^{2+} and nearly ideal tetrahedral $[BH_4]^-$. Each Mn^{2+} is surrounded by six octahedrally coordinated $[BH_4]^-$ anions and each $[BH_4]^-$ has three Mn^{2+} neighbors. Therefore this structure satisfies the electrostatic valence rule. The shortest Mn-H bond distance is 2.02 Å and the B-B nearest neighbor distance is 3.345 Å and Mn-B distance is 2.464 Å. The B-H bond lengths and the H-B-H bond angles are $d_{B\text{-}H}$ = 1.228 - 1.229 Å and $\theta_{H\text{-}B\text{-}H}$ = 106 – 112°, respectively. The geometry of the BH_4 tetrahedron is similar to that found in isolated

$[BH_4]^-$ ions, an ideal tetrahedral structure with a constant bond length of 1.24 Å for B-H bonds, and tetrahedral angles of 109.5°, where the B and H atoms are covalently bonded.

From electronic structure calculations, ionic interaction between Mn and $[BH_4]^-$ and strong ionocovalent B–H interaction within the $[BH_4]^-$ tetrahedral structure are revealed. Electronic density of states studies reveal that this phase has no band-gap which suggests that the $Mn(BH_4)_2$ is a metallic hydrogen storage material.

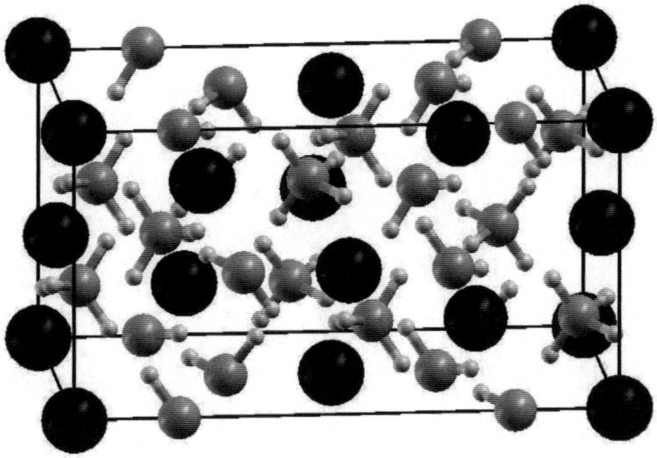

Figure 1. Crystal structure of $Mn(BH_4)_2$. Black (large), blue (middle) and green (small) spheres represent Mn, B and H atoms, respectively.

Conclusion

We have used density functional theory and lattice dynamics calculations to predict the crystal structure of the candidate material, the reaction enthalpies and identify suitable dehydrogenation reaction thermodynamics. It was found that $Mn(BH_4)_2$ exists as an orthorhombic structure of space group of *Fddd*. The electronic density of states shows that manganese borohydride is a metallic hydrogen storage material without having any band gap between valence and conduction bands. The electronic structure analysis also implies polar covalent bonds both between Mn and H and between B and H, with the lowest degree of polarity between Mn and H. The enthalpy of formation was calculated for the $Mn(BH_4)_2$ to be -135.026 kJ/f.u. at 0 K including zero point energy (ZPE) corrections. The simulated standard enthalpy of formation and dehydrogenation reaction for the complex $Mn(BH_4)_2$ shows that the most favorable dehydrogenation reaction is found to be $Mn(BH_4)_2 = Mn + 2B + 4H_2$, which is an endothermic reaction.

Acknowledgement

We gratefully acknowledge the US Department of Energy (Hydrogen Fuel Initiative code: DE-FG36-04G014224) and the National Science Foundation (award CHE-0722887) for funding this work. We also thank Academic Computing at the University of South Florida for providing computational support.

Reference

1. Schlapbach, L. and Zuttel, A., Nature **414**, 353 (2001).
2. Grochala, W. and Edwards, P. P., Chem. Rev. **104**, 1283 (2004).
3. G. Kresse and J. Furthmüller, Phys. Rev. B **54**, 11169 (1996).
4. G. Kresse and J. Furthmüller, Comput. Mater. Sci. **6**, 15 (1996).
5. G. Kresse and J. Hafner, Phys. Rev. B **47**, R558 (1993).
6. Choudhury, P., Bhethanabotla, V. R., and Stefanakos, E., Physical Review B **77**, 134302 (2008).
7. K. Parlinski, Software PHONON (2005)

102d Modeling Sustained Energetic Chain Reactions In Carbon Nanotubes

Nitish Nair, Department of Chemical Engineering, Massachusetts Institute of Technology, Building 66-580, 25 Ames St, Cambridge, MA 02141 and Michael S. Strano, 66-566 Department of Chemical Engineering, Massachusetts Institute of Technology, 77 Massachusetts Avenue, Cambridge, MA 02139.

We have developed a computational technique for tracking chain reactions along functionalised one-dimensional lattices, specifically single walled carbon nanotubes (SWNT). A hybrid Molecular Dynamics/Monte Carlo algorithm has been used for a coarse-grained model of a SWNT. The parametric space comprises the activation energies of the functional groups, and the fraction of enthalpy released during the reaction which is converted into translational kinetic energy. With the help of these two handles, we have estimated the stabilities of the proposed nanostructures and explored boundaries within which a reaction that has been initiated at one end of the nanotube will be sustained along its length. Finally, the effect of the density of loading of the functional groups has been studied with respect to the aforementioned details.

102e Block Copolymer Electrolytes for Flexible Thin-Film Lithium Batteries

Ayan Ghosh, Department of Chemical and Biomolecular Engineering, University of Maryland, 2113 Chemical and Nuclear Engineering Bldg, College Park, MD 20742 and Peter Kofinas, Fischell Department of Bioengineering, University of Maryland, 1120 Jeong H Kim Engineering Bldg 225, College Park, MD 20742.

As technology evolves, leading to a constant miniaturization of electrical devices, the need for a compact, lighter, safer and versatile energy source has been a critical limiting factor. Owing to its highly reactive nature, lithium batteries currently in use require thick and bulky containment layers in order to protect its components. Polymer electrolytes offer an attractive alternative by eliminating the safety risks posed by the corrosive organic electrolytes currently in use and also by being light-weight and flexible. Poly(ethylene oxide) (PEO) a popular solvent-free electrolyte, has been known for its ability to complex with lithium based salts. It however, suffers from low conductivity performance at room temperature owning to its semi-crystalline nature. We have developed a solid state nanostructured electrolyte based on a self-assembled block copolymer, which exhibits enhanced lithium transport. The electrolyte forms flexible translucent films which exhibit an average ionic conductivity value of 1.26×10^{-5} Scm^{-1} at room temperature (21 °C). Conductivity data was obtained from impedance response, while differential scanning calorimetry (DSC) and transmission electron microscopy (TEM) were performed to characterize the intrinsic properties of the material.

102f Computational Study of Electrical Energy Storage In Supercapacitors Based on Carbon Nanotube Forests

Lu Yang[1], Lawrence R. Pratt[2], Brian H. Fishbine[1], David E. Hanson[1], and Albert Migliori[1]. (1) T-12, Los Alamos National Laboratory, Los Alamos, NM 87545, (2) Department of Chemical and Biomolecular Engineering, Tulane University, Room 334, Lindy Boggs Center, New Orleans, LA 70118

Depletion of fossil fuels, increased energy consumption, and the desirability of low CO2 emissions have heightened the need for efficient, clean, and renewable energy sources. However, many carbon-neutral renewable energy resources, e.g. solar and wind, are intermittent and require effective energy storage. Achieving a significant increase in the energy density of capacitive energy storage devices, while retaining their characteristic advantages of high power and extraordinary cyclability will open up great possibilities for the use of these devices across the entire energy sector. At present, there is a lack of fundamental understanding of the molecular interactions in both bulk electrolyte and the electrode-electrolyte interface, which has lead to an Edisonian approach to improvements in these systems. Here I will describe the first molecular scale calculations on the performance of supercapacitors based on carbon nanotube forests as electrodes, systems proposed to respond to these effective energy storage needs. To provide baseline information for modeling the capacitance of nanotube forests, we first calculated the dielectric constant of propylene carbonate, both as a uniform bulk liquid and in thin films confined between oppositely charged parallel graphite electrodes. We then calculated the capacitance for a fully realistic simulation of the experimental system. We find that the computational results are in good agreement with experiment. Calculations for different spacings of the nanotubes confirm an anomalous dependence of capacitance on pore radius that has been reported previously for supercapacitors. The present results therefore offer a basis for better understanding of capacitive energy storage and provide new insight into device design.

Polymer based carbons as potential materials for energy storage in Lithium /Oxygen batteries

Mojtaba Mirzaeian [a] ***and Peter J Hall*** [b]

Department of Chemical and Process Engineering, University of Strathclyde, Glasgow, G1 1XJ, Scotland, UK

Abstract: Porous carbon aerogels are prepared by polycondensation of resorcinol (R) and formaldehyde (F) catalyzed by sodium carbonate (C) followed by carbonization of the resultant aerogels at 800 °C in an inert atmosphere. The porous texture of the carbons has been adjusted by the change of the molar ratio of resorcinol to catalyst (R/C) in the gel precursors in the range of 100 to 500. The porous structure of the aerogels and carbon aerogels are characterized by N_2 adsorption-desorption measurements at 77 K. It is found that total pore volume and average pore diameter of the carbons increase with increase in the R/C ratio of the gel precursors. The prepared carbon aerogels are used as active materials in fabrication of composite carbon electrodes. The electrochemical performance of the electrodes has been tested by using them as cathodes in a Li/O_2 cell. Through the galvanostatic charge/discharge measurements, it is found that with an increase of R/C ratio, the specific capacity of the Li/O_2 cell fabricated from the carbon aerogels increases from 716 to 2077 mA h g^{-1}. The resulting voltage profiles for the first ten charge/discharge cycles indicate that the carbon samples possess excellent stability on cycling.

Key words: Rechargeable battery, lithium, oxygen, carbon, porosity, capacity.

[a] E-mail address: mojtaba.mirzaeian@strath.ac.uk ; Tel: +44 -141-5482075

[b] E-mail address: p.j.hall@strath.ac.uk ; Tel: +44 -141-5484084

1 Introduction

Increasingly global warming due to carbon emission is causing an energy revolution by requiring fossil fuels to be supplemented by alternative energy sources and by requiring changes in lifestyle. Thus the future energy landscape foresees a reduction in the use of fossil fuels by exploiting renewable energies as key energy supplies. Addressed to this need for environmental protection, there is extensive interest in the use of renewable energies and development of improved methods for energy storage to ensure the continuity of supply. To satisfy the critical demand for energy storage, it is essential to develop storage technologies that are capable of storing and delivering energy with longer service life as well as higher energy and power densities.

Due to their high energy density, high specific energy and high operating voltage, rechargeable lithium batteries are expected as the potential power sources for many applications such as consumer electronics and electric vehicles (EV) [1]. Lithium is an extremely light metal and has the lowest electronegativity of the standard cell potential -3.045 V in the existing metals [2]. It can donate electrons the most easily to form positive ions [3]. Commercial rechargeable lithium batteries use lithium transition metal oxides, typically $LiCoO_2$, as cathode and graphite as anode [4]. Their specific capacity depends on the capacities of both the cathode and anode materials. Graphite has a theoretical capacity of 372 mA h g^{-1} [5]. Energy storage in these batteries is limited by the cathode and does not exceed 200 mA h g^{-1} [6].

The capacity of a lithium battery system can be enhanced remarkably by using a completely different approach which combines Li as anode directly with oxygen as cathode active material in a Li/oxygen cell [6, 7]. Oxygen accessed from environment is reduced catalytically on an air electrode surface to form either an oxide or peroxide ion.

The catalytically formed anions react with lithium cations supplied by the anode and delivered by the electrolyte to form Li_2O_2 on the air electrode surface during discharge process. Oxygen the unlimited cathode active material is not stored in the battery and therefore the theoretical specific energy of the lithium/ oxygen battery (11400 W h kg^{-1} excluded oxygen and 5200 W h kg^{-1} included oxygen) is the highest among various metal/oxygen batteries and is still much larger than that achievable using a conventional lithium ion battery [7].

The O_2 electrode in lithium/oxygen batteries is a carbon having a porous structure in which several electrochemical and transport processes occur simultaneously. The porous structure acts as gas transport pores for the diffusion of oxygen to the carbon-electrolyte interface (reaction zoon), formation and storage of Li_2O_2 during the discharge process and also electrochemical decomposition of Li_2O_2 during the charge process. The Kinetics of oxygen diffusion through the cathode limits the performance of the Li/oxygen cells [8]. It has been also discussed that the end-of-discharge of the cell is reached when the carbon pores are filled or choked by the deposition of Li_2O_2 [6, 9]. This shows that the cell performance strongly depends on the morphology and structure of the carbon. Therefore the main challenge in this issue is the development of new carbon electrode materials to improve the kinetics of the air cathode and enhance the capacity, energy and power densities and the stability of the energy delivered by these systems.

In this work we aim to prepare high capacity electrodes based on R-F carbon aerogels as cathode electrode for Li/O_2 batteries. The porous structures of the carbons are discussed with emphasize on the synthesis routes and porosity and surface area measurements. The electrochemical performance of the carbon aerogel electrodes are examined by galvanostatic charge / discharge tests using a Solartron 1470 equipment.

2 Experimental

2.1 Chemicals

Resorcinol [$C_6H_4(OH)_2$, Sigma- Aldrich, 98%], formaldehyde [(HCHO), Sigma- Aldrich, 37% solution], sodium carbonate anhydrous [Na_2CO_3, Sigma- Aldrich, $\geq$ 99.5%], acetone [Fisher Scientific, reagent grade], lithium hexafluorophosphate [$LiPF_6$, Sigma- Aldrich, $\geq$ 99.99 % (battery grade)], lithium ribbon [Li, Sigma- Aldrich, 99.9%], Propylene carbonate anhydrous [$C_4H_6O_3$, Sigma- Aldrich, 99.7%]

2.2 R-F aerogel synthesis

R-F hydrogels were synthesized by polycondensation of resorcinol [$C_6H_4(OH)_2$] (R) with formaldehyde [HCHO] (F) according to the method proposed by Pekala et al [10,11]. R-F solutions were prepared by mixing the required amount of resorcinol (R), formaldehyde (F), sodium carbonate (C) and distilled water (W) under vigorous stirring for 45 minutes. The homogeneous R-F solutions were poured into sealed glass vials to prevent the evaporation of water during gelation and gelled by aging for 72 hours at 90 °C into an oven to obtain R-F hydrogels. The hydrogels have a structure filled with water. To remove water from their structure, the gels were solvent exchanged by immersing into acetone for several hours prior to the vacuum drying. Finally the hydrogels were dried in a vacuum oven at 80 °C for 3 days. R-F aerogels with different resorcinol to catalyst (R/C) ratios were prepared by this method.

2.3 Preparation of carbon aerogels

Carbonization of the resultant R-F aerogels was performed at 800 °C in a tubular furnace using the following method: Typically 3 grams of gel was used in each carbonization experiment. The gel was placed in a ceramic boat and purge with Ar flowing at 200 ml min^{-1} for 30 min at room temperature prior to the heating program. The furnace was heated to 150 °C at 5 °C min^{-1} and maintained at 150 °C for 30 min. Then it was heated to 400 °C at 5 °C min^{-1} and hold for 30 min. Finally it was heated to the carbonization temperature at 10 °C min^{-1} and kept for 3 hours. The whole heating program was performed under Ar flowing at 200 ml min^{-1}. After pyrolysis, the furnace was cooled in flowing Ar to room temperature.

2.4 Material characterization

The pore texture of dried gel and carbon materials were studied by the analysis of nitrogen adsorption-desorption isotherms measured by an ASAP 2000 adsorption analyzer (Micromeritics) at 77 K. The samples were evacuated at 373 K for 2 h prior to the adsorption measurements. BET method was used for surface area measurements, BJH adsorptions-desorption was used for mesopore analysis and t-plot method was used for micropore analysis. Total pore volume was calculated from the adsorbed volume of nitrogen at $P/P_0 = 0.99$ (saturation pressure). Pore size distributions were obtained by the BJH method from adsorption branch of the isotherms.

2.5 Cell construction and electrochemical measurements

Cathode electrodes were prepared by mixing porous carbon-aerogel sample, electrolytic manganese dioxide (EMD), Kynar Flex 2801 binder (polyvinylidene difluoride/hexafluoropropylene; PVDF/HFP 88/12 by weight) and propylene carbonate (PC) (weight ratios: 11/19/15/55) according to the procedure described elsewhere [7,9]. The materials are placed in a glass bottle to which acetone and a stirrer bar are added and the resulting mixture was left to stir for 4 hours to form a paste. Then the paste was spread into a 200 μm thick film by an applicator. The electrode was untouched until the acetone evaporates. Disks of 1.3 cm diameter were then cut to make disk-like cathode electrodes.

Electrochemical cell was constructed in an argon-filled glove box. The cell consists of a stainless still bar as anode current collector, a lithium metal foil as anode, a glass micro-fiber separator soaked with electrolyte (1 M LiPF6 in propylene carbonate), a cathode electrode, an aluminum mesh and a spring and a hollow aluminum bar on the top of it as cathode current collector. The cell parts were compressed together to ensure that the electrode contact is good and then the cell was completely sealed except for the Al mesh window that exposes the porous cathode to the O_2 atmosphere. The cell then was exposed to O_2 at atmospheric pressure for a minimum of 30 min prior to the electrochemical tests and all measurements were carried out under O_2 atmosphere at room temperature.

Galvanostatic charge/discharge tests were performed by a Solartron 1470 potentiostat by applying a constant current at charge/discharge rate of 50 mA g^{-1} in the voltage range between 2 and 4.5 V.

3 Results and discussion

3.1 Surface characterization of aerogel samples

Adsorption-desorption isotherms of aerogels prepared at various R/C ratios are shown in Figure 1. The adsorption-desorption isotherms for sample having R/C ratio of 200 show significant hysteresis loop with type-IV isotherm which is representative of the filling and emptying of mesopores in the P/P_0 region of 0.4 - 0.85 by capillary condensation of the adsorbate in the mesopores of the solid [12]. This suggests the mesoporosity in the solid [12, 13]. The lower part of the hysteresis loop represents the filling of the mesopores while the upper part represents the emptying of the mesopores.

In the case of RF003 sample with R/C = 500 the hysteresis loop is shifted to the higher relative pressures (i.e. P/P_0 region of 0.75 - 0.99) indicating the emptying and filling of wider pores. For RF001 sample with R/C = 100 adsorption-desorption isotherms show a small hysteresis loop in the P/P_0 region of 0.4 - 0.6.

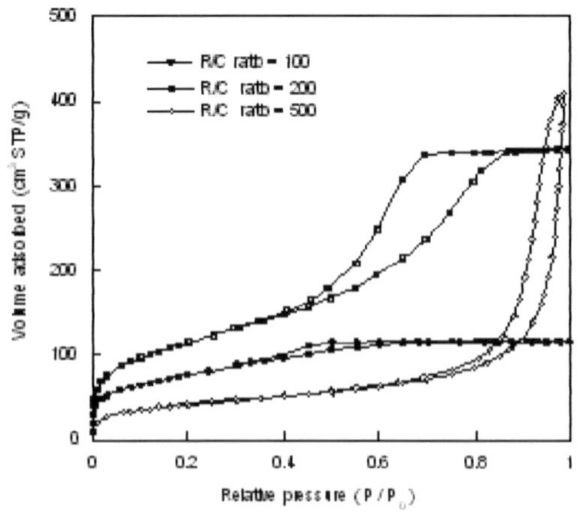

Figure 1. N_2 adsorption-desorption isotherms at 77 K for R-F aerogels prepared at various R/C ratios.

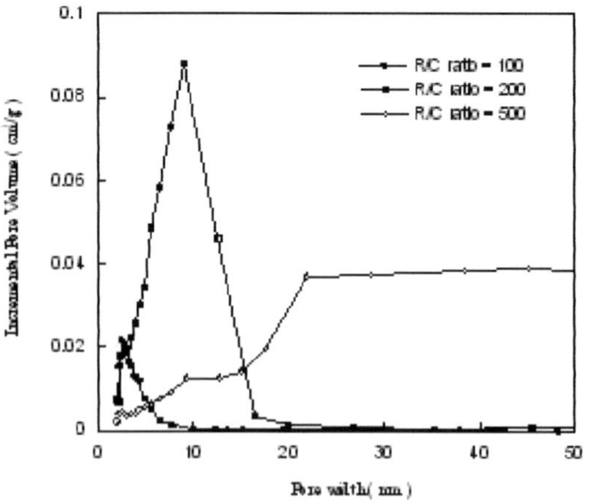

Figure 2. Pore size distribution of R-F aerogels with different R/C ratios.

Table 1. Surface characterization of R-F aerogel samples prepared at various R/C ratios

Sample	R/F	R/C	R/W	S_{BET} m^2/g	V_{total} cm^3/g	V_{micro} cm^3/g	V_{meso} cm^3/g	$\%V_{micro}$	$\%V_{meso}$	D_{avg} nm
RF001	0.50	100	0.1	278	0.1797	0.1189	0.0608	66.16	33.83	2.58
RF002	0.50	200	0.1	417	0.5301	0.1801	0.350	33.44	66.56	5.10
RF003	0.50	500	0.1	149	0.632	0.0075	0.6245	1.186	98.814	16.96

From data for the isotherms of the R-F aerogels, surface area, pore volume, and pore size distribution were calculated. The pore size distribution was calculated using the BJH method. The micro pore volume was calculated using t-plot method and mesopore volume was determined by difference. The above mentioned characteristics of the R-F aerogels are shown in Figure 2 and Table1. Data given in Table 1 show that increasing R/C ratio, increases total pore volume and level of mesoporosity in aerogels.

3.2 Effect of carbonization on the porous structure of the gels

Figure 3 shows adsorption and desorption isotherms of nitrogen on aerogel RF001 and carbon-aerogels CRF001 prepared at different carbonization temperatures.

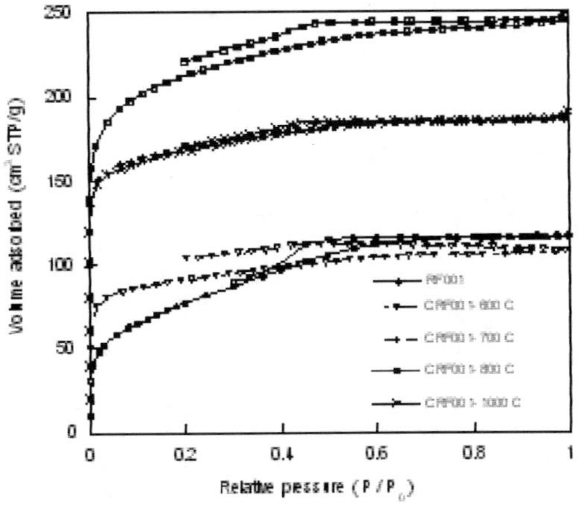

Figure 3. N_2 adsorption-desorption isotherms at 77 K for RF001 aerogel and CRF001 carbon-aerogels.

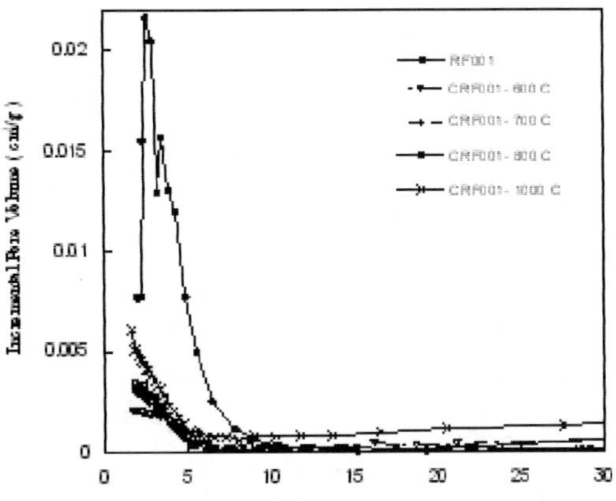

Figure 4. Pore size distribution of RF001 aerogel and CRF001 carbon-aerogels.

This figure suggests that the amounts of nitrogen adsorbed by carbon aerogels increases with increase in the carbonization temperature for T ≤ 800 °C. The results mean that porosity of carbon gels increases with increasing temperature in this range. As carbonization temperature increases to 800 °C, carbonization modifies the porous structure of the original gel and creates more micropores in the structure. This can be observed by a sharp rise at the initial region of the isotherm with temperature (i.e. at low relative pressures) which is indicative of the presence of micropores in carbon aerogels.

Figure 3 also shows that for carbonization temperatures above 800 °C, the amounts of nitrogen adsorbed by carbon aerogels decreases with increase in temperature which means decrease in the porosity of the carbon aerogel with increase in the carbonization temperature probably due to the collapse of the porous structure at high temperatures. Changes in the pore parameters of the carbon aerogels with heat treatment temperatures are listed in Table 2. Data shows that both total surface area and total pore volume increase with carbonization temperature for carbon aerogels below 800 °C. Heat treatment of carbon aerogel above 800 °C was found to decrease both total surface area and pore volume.

Table 2. Surface characterization of the aerogel and carbon aerogels prepared at various temperaments

Sample	Temperature (°C)	S_{BET} m^2/g	V_{total} cm^3/g	V_{micro} cm^3/g	V_{meso} cm^3/g	$\%V_{micro}$	$\%V_{meso}$	D_{avg} nm
RF001	-	278	0.1797	0.1189	0.0608	66	34	2.5830
CRF001-600	600	286	0.1870	0.1484	0.0386	79	21	2.3351
CRF001-700	700	524	0.2874	0.2625	0.0249	91	9	2.2708
CRF001-800	800	669	0.3800	0.3300	0.0500	86	14	2.1938
CRF001-1000	1000	520	0.2905	0.2603	0.0302	89	11	2.2316

Figure 4 shows the pore size distributions of the aerogel and carbon aerogels prepared at different temperatures. There is a significant change in mesopore size distribution of the gel after carbonization. Pore size distribution curves show that carbonization process decreases volume of pores significantly for pores with diameter in the range of 3 to 10 nm. Data given in Table 2 suggests that decrease in the mesoprosity of the gel is in expense of increase in microporosity during carbonization process as the total pore volume increases with heat treatment.

3.3 Surface characterization of carbon aerogels

Adsorption-desorption isotherms of nitrogen at 77 K for carbons prepared form aerogels with different R/C ratios at 800 °C are shown in Figure 5. The adsorption-desorption isotherms for CRF001 carbon aerogel (R/C = 100) show a type I isotherm according to IUPAC, indicating that this sample is microporous [12, 13]. The initial part of isotherm shows micropore filling and the low slop of the plateau is due to multilayer adsorption on the external surface.

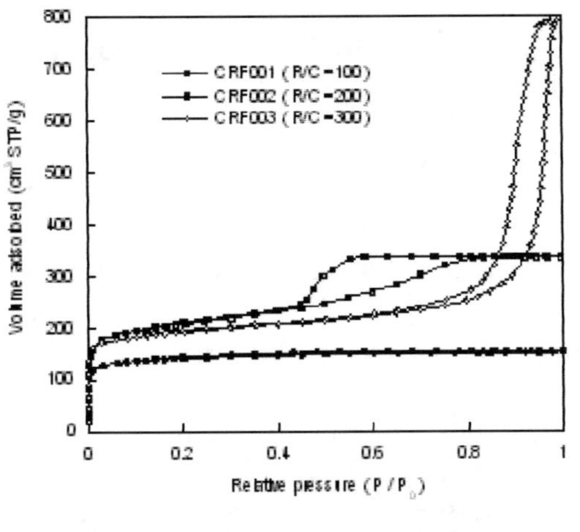

Figure 5. N_2 adsorption-desorption isotherms at 77 K for carbon-aerogels with different R/C ratios.

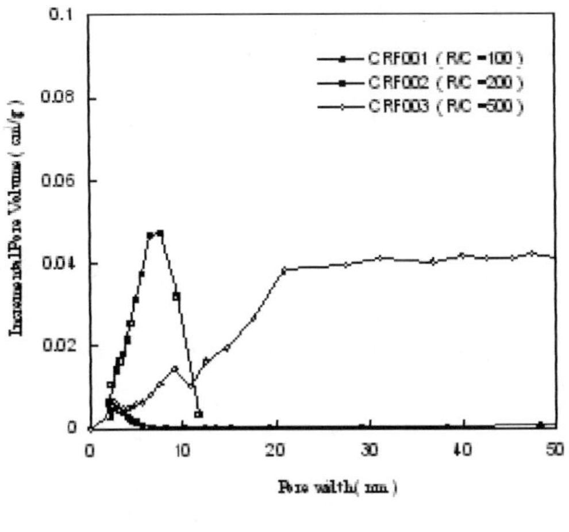

Figure 6. Pore size distribution of carbon-aerogels with different R/C ratios.

This shows that carbonization introduces large quantity of micropores into the carbon structure. For CRF002 sample (R/C = 200) the initial region of the isotherms (i.e. at low relative pressures: 0 – 0.2) experience a sharp rise which is indicative of the presence of micropores in the sample. Hysteresis loop in the pressure range 0.4 - 0.8 is indicative of mesoporosity in the structure of the carbon in addition to the presence of micropores. In the case of CRF003 sample (R/C = 500) hysteresis loop for filling and emptying of mesopores is shifted to a higher pressure range (i.e. 0.7 – 1.0) which is associated with the presence of mesopores with larger size in the structure. Pore size distribution of the carbon samples are shown in Figure 6. Sample CRF001 exhibits a micropore peak at the pore width of 2 nm and there is a shift of pore width to 7.8 nm and 20.9 nm for CRF002 and CRF003 samples respectively. The porous textural parameters, such as specific surface area, pore volume and average pore diameter for carbon samples are listed in Table 3.

Table 3. Porous textural parameters of carbon samples

Sample	R/C	S_{BET} m^2/g	V_{total} cm^3/g	V_{micro} cm^3/g	V_{meso} cm^3/g	$\%V_{micro}$	$\%V_{meso}$	D_{avg} nm
CRF001	100	669	0.3800	0.3300	0.0500	86.84	13.16	2.1938
CRF002	200	707	0.5270	0.1884	0.3386	35.75	64.25	4.4234
CRF003	500	647	1.2245	0.2101	1.0144	17.16	82.84	21.6503

It can be observed that with increase of the mole ratio of R/C in the initial gels both the total pore volume and the average pore diameter of the resultant carbons increase from 0.38 to1.2245 $cm^3\ g^{-1}$ and 2.1938 to 21.6503 nm, respectively. Three carbons prepared in this work possess different porous structures ranging from microporous structure, small mesopore in the structure and large mesopore in the structure. These results indicate that the porous texture and specially the pore volume and the pore size distribution of the carbon materials can be adjusted by changing the R/C ratio of the gel precursors. Pores in carbon electrode allow both the oxygen gas and electrolyte to come in contact with the catalyst and carbon. The electrochemical reduction of oxygen takes place on the carbon surface. Therefore it is important to minimize diffusion barriers to increase storage capacity of Li/O_2 cells by tailoring carbon electrodes having appropriate porous textures and pore volumes that maximize electrolyte and oxygen accessibility to the porous network. Electrochemical performance of cathode electrodes made from the above mentioned carbons are studied in order to find high capacity carbon based electrodes for lithium/oxygen cells.

3.4 Electrochemical studies

All the data reported below are obtained from the galvanostatic charge/discharge cycling of the cells assembled as described above. Assuming that the products of the discharge reaction in Li/O_2 cell deposit only onto carbon, the charge/discharge capacity of the lithium/oxygen cells has been normalized to the weight of carbon in the cathode electrode.

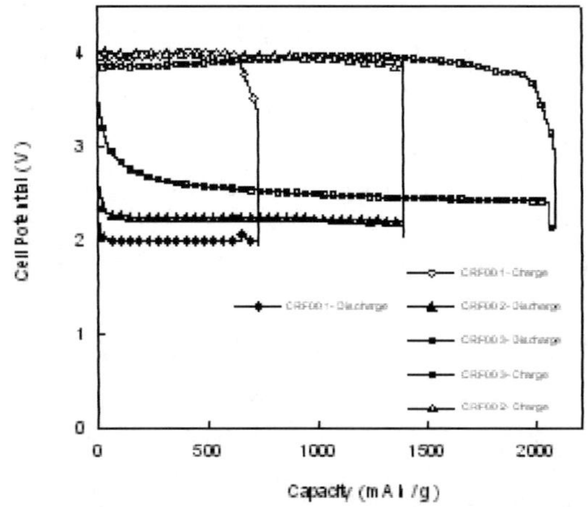

Figure 7. Discharge/charge behavior of the Li/O_2 cell at 50 mA g^{-1} based on electrodes prepared from CRF001, CRF002 and CRF003 carbons

Figure 7 shows a plot of charge/discharge characteristics of the Li/O_2 cells (third cycles) based on electrodes prepared from CRF001, CRF002 and CRF003 carbons. Variations of Cell voltage are plotted against specific capacity of the active material (carbon) when cells were charged and discharged at constant current rate of 50 mA g^{-1}.

Three carbon samples show flat discharge and charge profiles ranging between 2 to 2.5V and 3.85 to 4.05 respectively. These are associated with the storage behavior of the cells and are in good agreement with the preceding reported data for a similar cell discharged in O_2 at atmospheric pressure involving formation of Li_2O_2 as discharge product [6, 7]. Results of cyclic discharge/charge testing of carbon samples are given in Table 4.

Table 4. Results of galvanostatic discharge/charge testing of carbon samples

Sample	Charge voltage (V)	Discharge voltage (V)	Average reversible capacity (mA.h/g)
CRF001	3.97	2.0	716
CRF002	4.05	2.2	1381
CRF003	3.85	2.5	2077

Data given in Table 4 show that type of carbon used in cathode electrode has a significant effect on discharge capacity at 50 mA g^{-1}. CRF003 sample with the highest pore volume shows the highest discharge capacity (2077 mA h g^{-1}) while CRF001 sample with the lowest pore volume has the lowest discharge capacity (716 mA h g^{-1}) between carbon samples. The discharge reaction at the cathode results in the deposition of Li_2O_2 or Li_2O on the surface and pores of the carbon based cathode. Discharge ends when the cathode is choked off with deposit and all of the pores in the electrode are filled [6]. It is presumed that porous structure of carbon is the main factor controlling termination of discharge process and affecting the capacity of the cell. As discharge process proceeds, Li_2O_2 forms and fills pores in the carbon structure. If the porous structure is dominated by micropores, formation of discharge products chocks off micropores and significant fraction of the porous structure may not be electrochemically accessible which results in a small capacity for the cell. This is what has happened in a cell using CRF001 sample as active material in the cathode. When the R/C ratio increases to 500, the average reversible capacity increases to 2077 mA h g^{-1} the highest value in all samples. The large discharge capacity of CRF003 carbon might be due to the appropriate fraction of mesopres formed during the gelation and carbonization processes.

Mesopores are important for improving the performance of Li/O_2 cells. They lead the electrolyte ions transport into the bulk of material quickly and provide larger space for formation of discharge products. Therefore for a sample with wider pores and larger pore volume, polarization due to the formation of solid Li_2O_2 in pores is less and consequently larger discharge capacity is obtained.

Associating the reversible capacity with the textural parameters of the porous carbon materials shows that appropriate pore volume and average pore diameter are key factors contributing to high capacity. This increased capacity might be due to the contributions of higher diffusivity of the electrolyte into the carbon structure which results in the better transport and accessibility of lithium ions to the carbon surface (reaction zoon), larger space for the storage of discharge products as larger pore volume is available in the porous structure, and finally better diffusion of oxygen to the carbon-electrolyte interface. Change in the conductivity of the cathode with its pore volume is an important issue in the performance of Li/O_2 batteries which needs to be studied.

One of the vital characteristics of a rechargeable cell is the ability to accept repeated charge/discharge cycles without significant performance deterioration. The resulting voltage profiles for the first ten charge/discharge cycles for cycling of Li/O_2 cells using CRF001 and CRF003 as active materials between 2 and 4.5 V at 50 mA g^{-1} are shown in Figures 8-9.

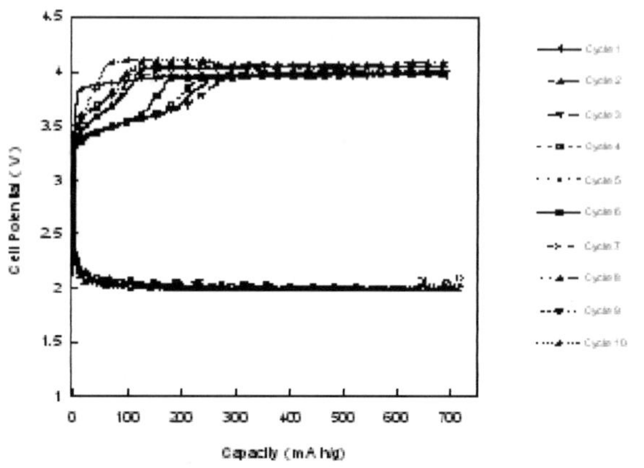

Figure 8. First ten charge/discharge curves for CRF001 carbon at 50 mA g^{-1}.

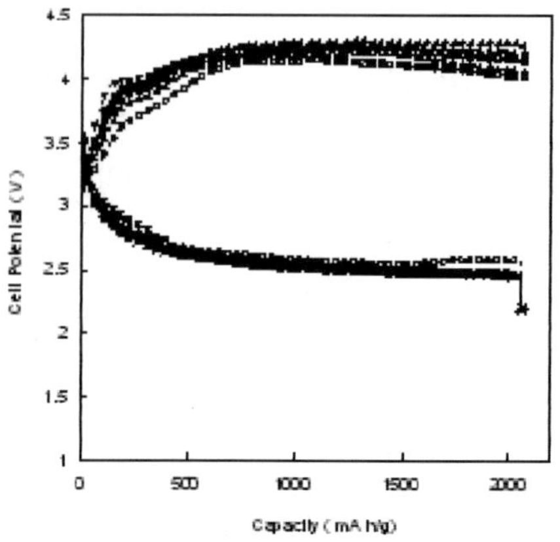

Figure 9. First ten charge/discharge curves for CRF003 carbon at 50 mA g^{-1}.

Charge/discharge cycles show that CRF001 and CRF003 based electrodes possess good cycling efficiencies and specific reversible capacities at these current density and voltage range.

4 Conclusions

Porous R-F gels have been synthesized through polycondensation of resorcinol with formaldehyde catalyzed by sodium carbonate. The porosity of the final gel is adjusted by changing the molar ratio of resorcinol to formaldehyde in the range of 100 to 500. The gels are used as precursors to prepare porous carbon materials with controlled porous structure by carbonization at 800 °C in an inert atmosphere. With an increase of R/C ratio from 100 to 500, both the total pore volume and average pore diameter of the carbons increased from 0.38 to1.2245 cm^3 g^{-1} and 2.1938 to 21.6503 nm, respectively. These porous carbon materials are investigated as active materials in cathode electrodes for Li/O_2 batteries and their electrochemical behavior has been characterized by constant current charge/discharge experiments. The results of electrochemical measurements show that with an increase of R/C ratio in carbon precursors from 100 to 500, the specific capacity of the Li/O_2 cell fabricated from the resultant carbons increases from 716 to 2077 mA h g^{-1}. The resulting voltage profiles for the first ten charge/discharge cycles indicate that the carbon samples sustain excellent stability on cycling.

Acknowledgments

We thank the EPSRC for funding as part of the Supergen Energy Storage Consortium (grant code EP/D031672/1).

References

[1] D. Linden, B. T. Reddy, Hand book of batteries, 3rd ed., McGraw-Hill, New York, 1989.

[2] M. Endo, C. Kim, K. Nishimura, Fujino T, K. Miyashita, Carbon, 38 (2000) 183-197.

[3] JP. Gabano, Lithium batteries, Academic press, New York, 1983.

[4] S. Flandrois, B. Simon, Carbon, 37 (1999) 165-180.

[5] R. Yazami, Ph. Touzain, J. Power Sources, 9 (1983) 365.

[6] K. M. Abraham, Z. J. Jiang, J. Electrochem. Soc., 143 (1996) 1-5.

[7] T. Ogasawara, A. Debart, M. Holzapfel, P. Novak, P.G. Bruce, J. Am. Chem. Soc., 128 (2006)1390-1393.

[8] J. Read, K. Mutolo, M. Ervin, W. Behl, J. Wolfenstine, A. Driedger, D. foster, J. Electrochem. Soc., 150 (2003) A1351-A1356.

[9] J. Read, J. Electrochem. Soc., 149 (2002) A1190-A1195.

[10] R W. Pekala, J. Mat. Sci., 24 (1989) 322.

[11] R W. Pekala, W. Schaefer, Macromolecules, 26 (1993) 5487-5493.

[12] F. Rouquerol, J. Rouquerol, K. Sing, Adsorption by powders and porous solids, principles, methodology and applications, Academic Press, New York, 1999.

[13] S. J. Gregg, K. S. W. Sing, Adsorption, surface area and porosity, Academic Press, New York, 1967.

163b Preparation of SnO2-Carbon Composite Hollow Spheres and Their Application In Lithium Batteries

Xiong Wen (David) Lou, Division of Bioengineering, School of Chemical and Biomedical Engineering, Nanyang Technological University, Block N1.3 Level B5-01, 70 Nanyang Avenue, Singapore, 637457, Singapore and Lynden A. Archer, Chemical and Biomolecular Engineering, Cornell University, 120 Olin Hall, Ithaca, NY 14853.

We have developed a simple one-step template-free method for the controlled synthesis of hollow or hollow core/shell-type SnO_2 nanostructures with controllable sizes in the range of 200 – 500 nm. This method is based on hydrothermal treatment of stannate in a mixed ethanol-water solvent. An inside-out Ostwald ripening mechanism is proposed to account for the template-free formation of these hollow nanostructures. A shortcoming of the method is that size uniformity of the as-synthesized SnO_2 hollow nanoparticles is difficult to achieve. To solve this problem, the hydrothermal method has been extended to synthesize SnO_2 hollow nanostructures with controlled sizes using monodisperse silica nanospheres as templates. Remarkably, uniform surface coating can be achieved without any prior surface modification. By repeated hydrothermal deposition or controlling the synthetic conditions, multi-shelled structures can also be prepared. These hollow SnO_2 nanostructures may find applications in many important technological fields. Here we study their application as electrodes for lithium ion batteries and report encouraging initial results. We have also designed new nanostructured SnO_2/carbon composite anode materials with hollow spherical structure. 3D carbon networks or shells in nanoscale can be effectively integrated with the SnO_2 shells by selective infiltration of carbon precursor in solution followed by carbonization. The carbon networks act as both physical buffering cushion for the intrinsic large volume change and electrical conducting path. As a result, the capacity retention of the composite electrode is largely improved.

163c Fabrication of Nanoscale High-Order Hierarchical Sn / C Composites for Highly Reversible Li+ Ion Storage

Da Deng and **Jim Yang Lee**. Department of Chemical & Biomolecular Engineering, National University of Singapore, 10 Kent Ridge Crescent, Singapore, 119260, Singapore

The rapid progress in portable electronic devices in this information-rich era demands continuing improvement in the performance of the power sources, which are typically the lithium-ion batteries. One of general strategies is to reduce the active electrode materials to the nanoscale to increase the capacity and rate capability of the electrodes. This is a report of our recent studies on the fabrication of various high-order Sn-carbon composites as lithium-ion storage compounds for the anode of lithium-ion battery. They range from hollow core-shell mesospheres, to hierarchical Rambutan-like and chestnut-like nanostructures. The methods of preparation are remarkably simple, environment-friendly, low cost and easily scaled up for volume production. In particular, the chestnut-like nanocomposites fabricated directly on copper current collectors are binder-free and conducting-additive free. Their double roughness structure of nanohairs on mesospheres also impart to them superhydrophobic properties mimetic of the lotus effect. For the unique hierarchical Rambutan-like nanostructure, tin remains electrochemically active even after 200 cycles of charge and discharge, thereby allaying some of the previous concerns on the use of tin as a lithium storage compound. These high-order nanomaterials also show impressively high capacity, high rate capability, and high reversibility demonstrating their potential as carbon alternatives for the next generation lithium ion batteries. This presentation will focus on the synthesis methods and the possible mechanisms of the formation of these intriguing morphologies.

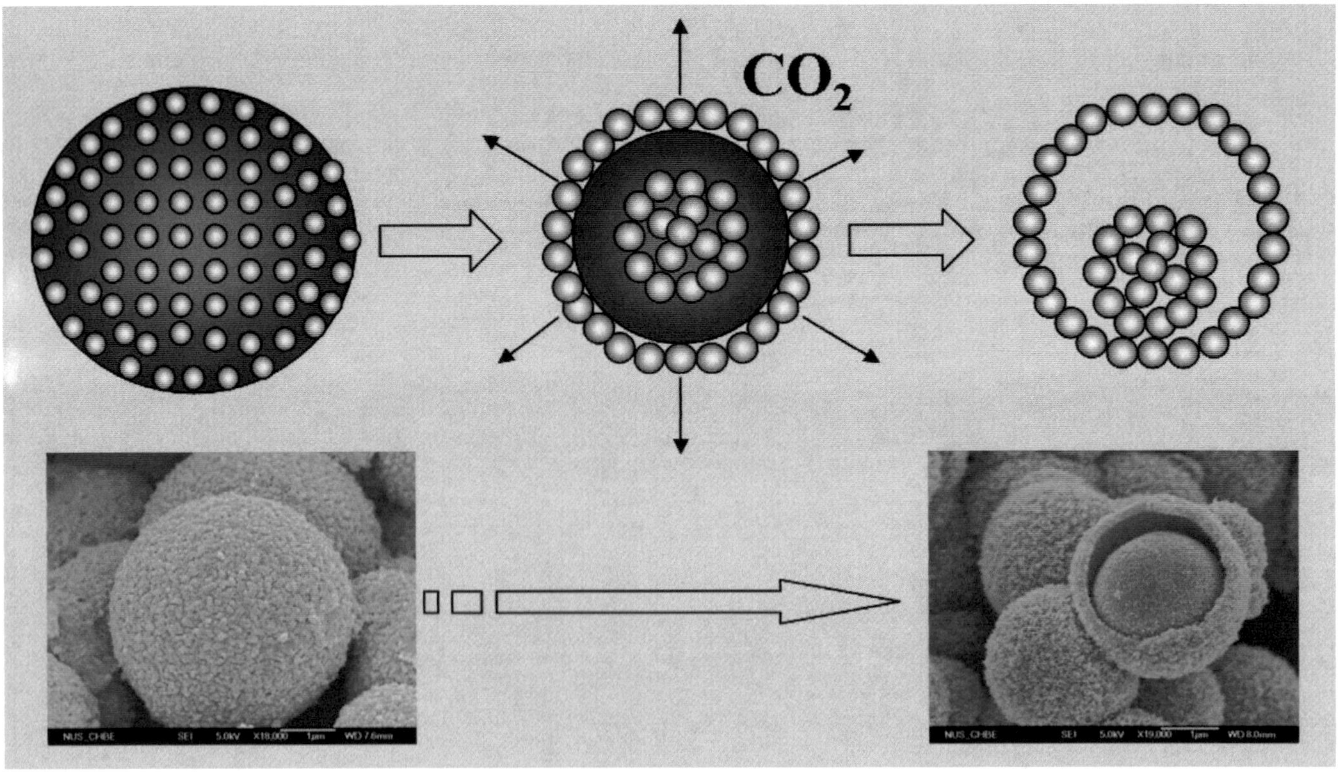

Web Page: www.chbe.nus.edu.sg/people/faculty_leejy.html

EVALUATION OF SILVER VANADIUM PHOSPHOROUS OXIDE AS A CATHODE MATERIAL IN LITHIUM PRIMARY CELLS

*Amy C. Marschilok[c], Kenneth J. Takeuchi[b], and Esther S. Takeuchi[a,b,c]**
a Department of Chemical and Biological Engineering
b Department of Chemistry
c Department of Electrical Engineering, University at Buffalo (SUNY), Buffalo, NY 14260

Introduction

Silver vanadium oxide (SVO, $Ag_2V_4O_{11}$) has drawn significant attention as the cathode material of choice in batteries used for implantable cardiac defibrillators.[1, 2] The application requires years of microampere current delivery with intermittent pulses in the ampere range. Lithium/SVO batteries successfully meet these demanding requirements. The discharge process of silver vanadium oxide in a lithium primary cell has been studied and begins with the reduction of Ag^+ to Ag^o followed by reduction of the vanadium centers from V(V) to V(IV) and V(III).[3] The reduced silver ions form silver metal particles and nanowires in the cathode matrix.[4] Consistent with the formation of the silver metal matrix, the conductivity of the cathode increases by several orders of magnitude. Similarly, copper vanadium oxides have also been investigated for their use as cathode materials in lithium based batteries, but in rechargeable systems.[5] The mechanism of reduction has been studied and also shown to progress with the formation of very small copper fibers or nano-wires on the surface of the electrode particles.[6]

Lithium iron phosphate, $LiFePO_4$ specifically,[7-53] and other phosphate based materials generally, have considered desirable for lithium battery cathode applications due to their to their high thermal and chemical stabilities.[54] However, a significant challenge to their practical implementation is overcoming their typically low electronic conductivity. Various methods to improve conductivity have been exploited including using carbon coatings,[13, 19, 55-73] doping agents,[12, 14, 18, 74-80] conductive current collectors,[52] the addition of conducting polymers.[81]

In order to obtain high stability and high conductivity characteristics, we have deliberately selected a silver vanadium phosphorous oxide cathode material. The layered phosphate based structure should possess an inherently low solubility and high stability in the battery cathode matrix. Similar to the silver vanadium oxide and copper vanadium oxide systems, the silver(I) ions in $Ag_2VO_2PO_4$ should be reduced to the metallic state via reduction elimination reactions as the batteries discharge, resulting in a significant increase in cathode pellet conductivity as a result of the reduction process. We propose therefore that silver vanadium phosphorous oxide will be an excellent cathode material for high power lithium battery applications.

Experimental

Silver vanadium phosphorous oxide ($Ag_2VO_2PO_4$) was prepared using a previously reported hydrothermal synthesis method.[82] $Ag_2VO_2PO_4$ was produced on a 0.5 g scale, by heating Ag_2O, V_2O_5, and H_3PO_4 in aqueous solution at 230°C in a Teflon-lined autoclave for 96 hours. The material was characterized by several methods including scanning electron microscopy (SEM), differential scanning calorimetry (DSC), and x-ray powder diffraction (XRD). A Micromeritics Tristar II 3020 was used for BET surface area and porosity analysis, with N_2 as the adsorbate. Samples were degassed under flowing $N_{2(g)}$ for 2 hours at 50°C prior to surface area analysis. A Micromeritics Accupyc II 1340 pycnometer was used for true density analysis made via helium gas displacement. A TA instruments Q600 was used for differential scanning calorimetry (DSC). A Rigaku Ultima IV x-ray powder diffractometer was used for XRD analysis. Cu Kα radiation was utilized at 40 kV, 44 mA, with Bragg-Brentano focusing geometry. MDI JADE version 8.5.3 software with ICDD and NIST databases was used for search-match analysis. Scanning electron microscopy (SEM) data was collected using a Hitachi SU-70 field emitting scanning electron microscope equipped with an Oxford Inca energy dispersive x-ray spectroscopy (EDS) system. Secondary electron images were acquired at 5 kV. Backscatter electron images were observed and EDS data was collected at

20 kV.

Cathode pellets were prepared by mixing Ketjenblack carbon, Fisher 38 graphite, powdered poly(tetrafluoroethane) (PTFE) binder and $Ag_2VO_2PO_4$, containing 79% $Ag_2VO_2PO_4$ by weight. Cathode pellets were pressed at 10 tons/cm^2. Type 2325 coin cells were fabricated within an Argon filled glove box. Electrolyte was 1 M $LiAsF_6$ in 50/50 (v/v) propylene carbonate/ dimethoxyethane. For the differential capacity test, coin cells were discharged at 37°C using a Maccor Series 4000 Battery Testing Unit. Cells were discharged under a constant current until the voltage dropped below 1.5 V. A 0.04 mA/cm^2 current was selected which allowed the cells to discharge completely within 12 days. For the pulse capability test, coin cells were discharged at room temperature and tested using a Bitrode SCN Cycle Life Tester battery testing unit. Four 10 second pulses at current densities of 20, 30, 40 and 50 mA/cm^2 were applied, separated by 15 second rests at open circuit potential. Between pulse trains, the cells were rested at open circuit potential for 30 minutes. This process was repeated until the pulse voltage dropped below 0.5 V.

Results and Discussion

Material characterization

The XRD pattern was recorded of the silver vanadium phosphorous oxide ($Ag_2VO_2PO_4$) sample prepared in our laboratories and showed a good match to that reported in the literature, confirming the successful preparation of the material.[83] It should be noted that silver vanadium phosphorous oxide ($Ag_2VO_2PO_4$) crystals have been reported to be in the monoclinic space group C2/m. The structure itself consists of layers of dimers of edge-sharing VO octahedra and PO tetrahedra, extending parallel to the (001) crystallographic plane with silver ions between the layers as illustrated in Figure 1.[83]

Figure 1. Structure of $Ag_2VO_2PO_4$

Differential scanning calorimetry (DSC) was used to evaluate the silver vanadium phosphorous oxide material purity. One major endotherm was noted at 535° with no other isotherms observed. The $Ag_2VO_2PO_4$ powder had a BET surface area of 0.98 ± 0.02 m^2/g, with an average pore width ~70 Angstroms. In order to determine the particle morphology and particle size, scanning electron microscopy (SEM) was recorded for samples of the hydrothermally prepared material and revealed our $Ag_2VO_2PO_4$ to have a structure consisting of micron sized bladed particles, as shown in Figure 2. The particles were acicular in appearance with an aspect ratio greater than twenty for many of the particles.

The samples were also imaged by SEM in backscatter electron mode, in order to obtain visual confirmation of sample homogeneity . No differences in contrast were observed in backscatter mode, indicating that the samples were uniform in terms of atomic number composition. Localized energy dispersive x-ray spectroscopy (EDS) was also utilized on two particles of differing morphology, shown in Figure 3. Both particles had Ag:V:P ratios of approximately 2:1:1, as shown in Table 1. This indicates that the composition of our $Ag_2VO_2PO_4$ material was uniform, irrespective of particle morphology. In addition to the silver, vanadium, phosphorous, and oxygen observed, carbon was also detected during EDS, likely from the double sided carbon tape used to support the $Ag_2VO_2PO_4$ powder during SEM and EDS analysis.

a) 1000x magnification b) 7000x magnification

Figure 2. Scanning electron microscopy of $Ag_2VO_2PO_4$

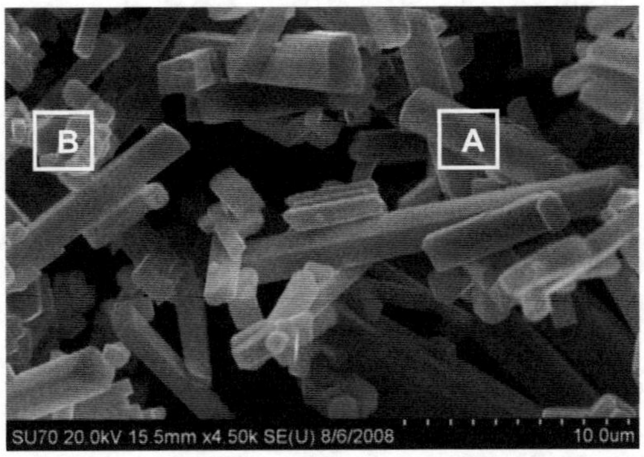

Figure 3. Scanning electron microscopy of $Ag_2VO_2PO_4$

Element	Atomic Composition (%)	
	Region A	Region B
Ag	13.8	9.4
V	7.0	4.3
P	8.4	4.9
C	31.4	33.6
O	39.4	47.8

Table 1. Energy dispersive x-ray spectroscopy analysis of $Ag_2VO_2PO_4$

Electrochemical evaluation

$Ag_2VO_2PO_4$ cathodes were prepared and placed in coin type cells containing lithium anodes. To our knowledge, this is the first use of $Ag_2VO_2PO_4$ as a cathode material. The voltage shows a gradual decrease from an initial potential of 3.3 V to an end of test potential of 1.5 V, as shown in Figure 4. Several plateaus or inflection points are present in the discharge curve.

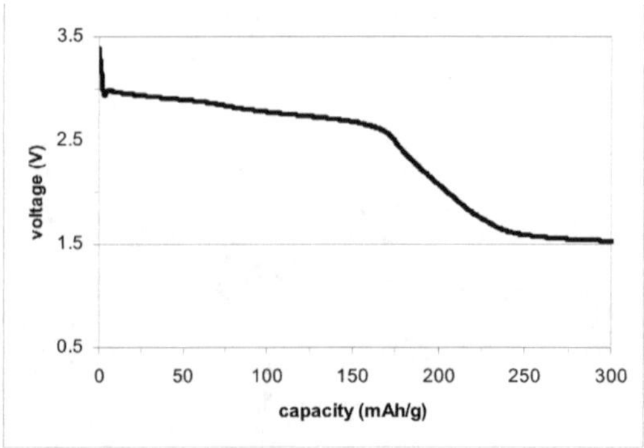

Figure 4. Voltage versus capacity for Li/$Ag_2VO_2PO_4$ batteries: constant current discharge test

The batteries delivered 270 mAh/g of $Ag_2VO_2PO_4$ cathode material above 1.5 V, as shown in Figure 4. Assuming complete reduction of Ag^+ to Ag^o and V^{+5} to V^{+3}, the $Ag_2VO_2PO_4$ would transfer 4 electrons per formula unit, translating to a 272 mAh/g theoretical gravimetric capacity. Notably, nearly 100% of the theoretical capacity of this material has been accessed. The measured true density of our $Ag_2VO_2PO_4$ was 5.32 g/cc. This translates to a high volumetric energy density of 1440 mAh/cc for $Ag_2VO_2PO_4$.

A second electrochemical test was conducted to assess the pulse power capability of the coin cells. Pulse trains consisting of four 10 second constant current pulses every 30 minutes. Each pulse of the train increased in current density and progressed as 20, 30, 40 and 50 mA/cm^2. The minimum voltage recorded under each pulse was recorded and is summarized in Figure 5. Even with a low surface area material and relatively large particle size, the cells demonstrated pulse capability of 50 mA/cm^2 above 1.5 V at the beginning of life. The pulse current under 30 mA/cm^2 remained above 1.5 volts through about 150 mAh/g of cathode material.

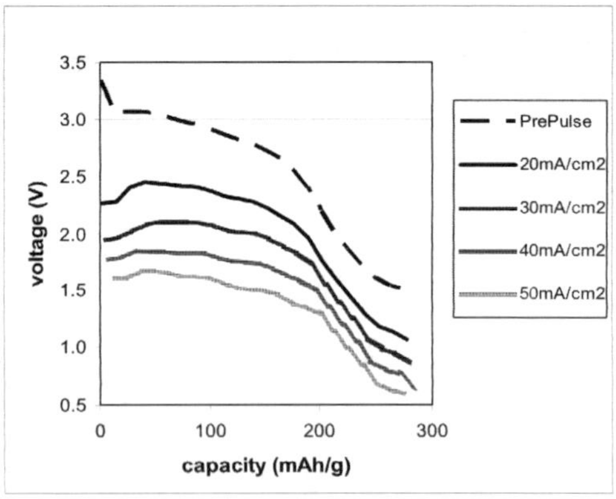

Figure 5. Voltage versus capacity of Li/$Ag_2VO_2PO_4$ batteries: pulse capability test

Summary

The silver vanadium phosphorous oxide $Ag_2VO_2PO_4$ has been established to be a viable cathode material for lithium primary cells. In addition, further research with this and additional members of the material family of mixed metal phosphorous oxides (*MM'*POs) holds promise for next generation high power battery applications.

References

1. Takeuchi, Kenneth J., Amy C. Marschilok, Steven M. Davis, Randolph A. Leising and Esther S. Takeuchi, (2001), "Silver Vanadium Oxides and Related Battery Applications," *Coord. Chem. Rev.*, 219, pp. 283-310.

2. Takeuchi, Kenneth J., Amy C. Marschilok and Esther S. Takeuchi, (2007), in *Vanadium: Chemistry, Biochemistry, Pharmacology and Practical Applications*, Alan S. Tracey, Gail R. Willsky and Esther S. Takeuchi Editors, Taylor and Francis, New York.

3. Leising, Randolph A., William C. Thiebolt, III and Esther Sans Takeuchi, (1994), "Solid-State Characterization of Reduced Silver Vanadium Oxide from the Li/Svo Discharge Reaction," *Inorg. Chem.*, 33, pp. 5733-5740.

4. Nancy R. Gleason, Randolph A. Leising, Marcus Palazzo, Esther S. Takeuchi, Kenneth J. Takeuchi, (Friday, October 1 2005), "Talk #248 - Microscopic Study of the First Voltage Plateau in the Discharge of Svo and the Consequences on Electronic Conductivity," *208th Meeting of the Electrochemical Society. Session D2 - Rechargeable Lithium and Lithium-Ion BatteriesBattery/Energy Technology.*

5. Tarascon, J. M., C. Delacourt, A. S. Prakash, M. Morcrette, M. S. Hegde, C. Wurm and C. Masquelier, (2004), "Various Strategies to Tune the Ionic/Electronic Properties of Electrode Materials," *Dalton Transactions*, pp. 2988-2994.

6. M. Morcrette, P. Rozier, L. Dupont, E. Mugnier, L. Sannier, J. Galy, J.M. Tarascon, (2003), "A Reversible Copper Extrusion-Insertion Electrode for Rechargeable Li Batteries," *Nature Materials*, 2, pp. 755-761.

7. Andersson, Anna S., John O. Thomas, Beata Kalska and Lennart Haggstrom, (2000), "Thermal Stability of Lifepo4-Based Cathodes," *Electrochem. Solid-State Lett.*, 3, pp. 66-68.

8. Andersson, Anna S., Beata Kalska, Lennart Haggstrom and John O. Thomas, (2000), "Lithium Extraction/Insertion in Lifepo4: An X-Ray Diffraction and Mossbauer Spectroscopy Study," *Solid State Ionics*, 130, pp. 41-52.

9. Arnold, G., J. Garche, R. Hemmer, S. Strobele, C. Vogler and M. Wohlfahrt-Mehrens, (2003), "Fine-Particle Lithium Iron Phosphate Lifepo4 Synthesized by a New Low-Cost Aqueous Precipitation Technique," *Journal of Power Sources*, 119-121, pp. 247-251.

10. Chen, J. J. and M. S. Whittingham, (2006), "Hydrothermal Synthesis of Lithium Iron Phosphate," *Electrochem. Commun.*, 8, pp. 855-858.

11. Chen, Yike, Shigeto Okada and Jun-ichi Yamaki, (2004), "Preparation and Characterization of Lifepo4/Ag Composite for Li-Ion Batteries," *Composite Interfaces*, 11, pp. 277-283.

12. Chung, Sung-Yoon and Yet-Ming Chiang, (2003), "Microscale Measurements of the Electrical Conductivity of Doped Lifepo4," *Electrochem. Solid-State Lett.*, 6, pp. A278-A281.

13. Chung, Sung-Yoon, Jason T. Bloking and Yet-Ming Chiang, (2003), "On the Electronic Conductivity of Phospho-Olivines as Lithium Storage Electrodes," *Nat. Mater.*, 2, pp. 702-703.

14. Croce, F., A. D'Epifanio, J. Hassoun, A. Deptula, T. Olczac and B. Scrosati, (2002), "A Novel Concept for the Synthesis of an Improved Lifepo4 Lithium Battery Cathode," *Electrochem. Solid-State Lett.*, 5, pp. A47-A50.

15. Deb, Aniruddha, Uwe Bergmann, Elton J. Cairns and S. P. Cramer, (2004), "X-Ray Absorption Spectroscopy Study of the Lixfepo4 Cathode During Cycling Using a Novel Electrochemical in Situ Reaction Cell," *Journal of Synchrotron Radiation*, 11, pp. 497-504.

16. Delacourt, C., P. Poizot, S. Levasseur and C. Masquelier, (2006), "Size Effects on Carbon-Free Lifepo4 Powders," *Electrochemical And Solid State Letters*, 9, pp. A352-A355.

17. Delacourt, Charles, Philippe Poizot, Jean-Marie Tarascon and Christian Masquelier, (2005), "The Existence of a Temperature-Driven Solid Solution in Lixfepo4 for 0 < X < 1," *Nature Materials*, 4, pp. 254-260.

18. Delacourt, C., C. Wurm, L. Laffont, J. B. Leriche and C. Masquelier, (2006), "Electrochemical and Electrical Properties of Nb- and/or C-Containing Lifepo4 Composites," *Solid State Ionics*, 177, pp. 333-341.

19. Delacourt, Charles, Calin Wurm, Priscilla Reale, Mathieu Morcrette and Christian Masquelier, (2004), "Low Temperature Preparation of Optimized Phosphates for Li-Battery Applications," *Solid State Ionics*, 173, pp. 113-118.

20. Dokko, K., S. Koizumi and K. Kanamura, (2006), "Electrochemical Reactivity of Lifepo4 Prepared by Hydrothermal Method," *Chem. Lett.*, 35, pp. 338-339.

21. Ellis, B. L., W. R. M. Makahnouk, Y. Makimura, K. Toghill and L. F. Nazar, (2007), "A Multifunctional 3.5 V Iron-Based Phosphate Cathode for Rechargeable Batteries," *Nat. Mater.*, 6, pp. 749-753.

22. Fisher, Craig A. J. and M. Saiful Islam, (2008), "Surface Structures and Crystal Morphologies of Lifepo4. Relevance to Electrochemical Behaviour," *J. Mater. Chem.*, 18, pp. 1209-1215.

23. Franger, S., C. Bourbon and F. Le Cras, (2004), "Optimized Lithium Iron Phosphate for High-Rate Electrochemical Applications," *J. Electrochem. Soc.*, 151, pp. A1024-A1027.

24. Guo, Z. P., H. Liu, S. Bewlay, H. K. Liu and S. X. Dou, (2003), "A New Synthetic Method for Preparing Lifepo4 with Enhanced Electrochemical Performance," *Journal Of New Materials For Electrochemical Systems*, 6, pp. 259-262.

25. Higuchi, M., K. Katayama, Y. Azuma, M. Yukawa and M. Suhara, (2003), "Synthesis of Lifepo4 Cathode Material by Microwave Processing," *Journal of Power Sources*, 119, pp. 258-261.

26. Higuchi, Masashi, Toshiyuki Tsuruoka, Takashi Asaka, Takuya Mihara, Manabu Suhara, Keiichi Katayama and Yasuo Azuma, (2004), "Preparation of Lifepo4 as Cathode Material for a Lithium Ion Battery by Microwave Processing," *Key Engineering Materials*, 269, pp. 147-150.

27. Hong, Young-Sik, Kwang Sun Ryu, Yong Joon Park, Min Gyu Kim, Jay Min Lee and Soon Ho Chang, (2002), "Amorphous Fepo4 as 3 V Cathode Material for Lithium Secondary Batteries," *J. Mater. Chem.*, 12, pp. 1870-1874.

28. Iriyama, Y., M. Yokoyama, C. Yada, S. K. Jeong, I. Yamada, T. Abe, M. Inaba and Z. Ogumi, (2004), "Preparation of Lifepo4 Thin Films by Pulsed Laser Deposition and Their Electrochemical Properties," *Electrochemical And Solid State Letters*, 7, pp. A340-A342.

29. Kim, Hyung-Sun, Byung-Won Cho and Won-Il Cho, (2004), "Cycling Performance of Lifepo4 Cathode Material for Lithium Secondary Batteries," *Journal of Power Sources*,

132, pp. 235-239.

30. Myung, Seung-Taek, Shinichi Komaba, Koutarou Kurihara and Naoaki Kumagai, (2003), "Preparation of Lifepo4 as Lithium Intercalation Compound by Emulsion Drying Method," *Electrochemistry (Tokyo, Japan)*, 71, pp. 177-179.

31. Ni, J. F., G. Y. Su, H. H. Zhou and J. T. Chen, (2004), "Study of Limpo4 as Cathode Material for Lithium Ion Batteries," *Progress In Chemistry*, 16, pp. 554-560.

32. Okada, S. and J. Yamaki, (2004), "Iron-Based Cathodes/Anodes for Li-Ion and Post Li-Ion Batteries," *Journal Of Industrial And Engineering Chemistry*, 10, pp. 1104-1113.

33. Sauvage, F., E. Baudrin, M. Morcrette and J. M. Tarascon, (2004), "Pulsed Laser Deposition and Electrochemical Properties of Lifepo4 Thin Films," *Electrochem. Solid-State Lett.*, 7, pp. A15-A18.

34. Scaccia, Silvera, Maria Carewska, Pawel Wisniewski and Pier Paolo Prosini, (2003), "Morphological Investigation of Sub-Micron Fepo4 and Lifepo4 Particles for Rechargeable Lithium Batteries," *Mater. Res. Bull.*, 38, pp. 1155-1163.

35. Shi, Siqi, Chuying Ouyang, Zhihua Xiong, Lijun Liu, Zhaoxiang Wang, Hong Li, Ding-sheng Wang, Liquan Chen and Xuejie Huang, (2005), "First-Principles Investigation of the Structural, Magnetic, and Electronic Properties of Olivine Lifepo4," *Physical Review B: Condensed Matter and Materials Physics*, 71, pp. 144404/144401-144404/144406.

36. Shim, Joongpyo and Kathryn A. Striebel, (2003), "Cycling Performance of Low-Cost Lithium Ion Batteries with Natural Graphite and Lifepo4," *Journal of Power Sources*, 119-121, pp. 955-958.

37. Shi, Z. C. and Y. Yang, (2005), "Progress in Polyanion-Type Cathode Materials for Lithium Ion Batteries," *Progress In Chemistry*, 17, pp. 604-613.

38. Srinivasan, Venkat and John Newman, (2004), "Discharge Model for the Lithium Iron-Phosphate Electrode," *J. Electrochem. Soc.*, 151, pp. A1517-A1529.

39. Striebel, K., A. Guerfi, J. Shim, M. Armand, M. Gauthier and K. Zaghib, (2003), "Lifepo4/Gel/Natural Graphite Cells for the Batt Program," *Journal of Power Sources*, 119-121, pp. 951-954.

40. Striebel, Kathryn, Joongpyo Shim, Venkat Srinivasan and John Newman, (2005), "Comparison of Lifepo4 from Different Sources," *J. Electrochem. Soc.*, 152, pp. A664-A670.

41. Sylvain Franger, Frederic Le Cras, Carole Bourbon, Helene Rouault, (2003), "Comparison between Different $Lifepo_4$ Synthesis Routes and Their Influence on Its Physico-Chemical Properties," *Journal of Power Sources*, 119-121, pp. 252-257.

42. Tajimi, S., Y. Ikeda, K. Uematsu, K. Toda and M. Sato, (2004), "Enhanced Electrochemical Performance of Lifepo4 Prepared by Hydrothermal Reaction," *Solid State Ionics*, 175, pp. 287-290.

43. Takahashi, Masaya, Shin-ichi Tobishima, Koji Takei and Yoji Sakurai, (2002), "Reaction Behavior of Lifepo4 as a Cathode Material for Rechargeable Lithium Batteries," *Solid State Ionics*, 148, pp. 283-289.

44. Tang, Ping and N. A. W. Holzwarth, (2003), "Electronic Structure of Fepo4, Lifepo4, and Related Materials," *Physical Review B: Condensed Matter and Materials Physics*, 68, pp. 165107/165101-165107/165110.

45. Wang, Chunsheng and Jian Hong, (2007), "Ionic/Electronic Conducting Characteristics of Lifepo4 Cathode Materials. The Determining Factors for High Rate Performance," *Electrochem. Solid-State Lett.*, 10, pp. A65-A69.

46. Wang, Deyu, Hong Li, Zhaoxiang Wang, Xiaodong Wu, Yucheng Sun, Xuejie Huang and Liquan Chen, (2004), "New Solid-State Synthesis Routine and Mechanism for Lifepo4 Using Lif as Lithium Precursor," *Journal of Solid State Chemistry*, 177, pp. 4582-4587.

47. Xu, Yong-Nian, Sung-Yoon Chung, Jason T. Bloking, Yet-Ming Chiang and W. Y. Ching, (2004), "Electronic Structure and Electrical Conductivity of Undoped Lifepo4," *Electrochem. Solid-State Lett.*, 7, pp. A131-A134.

48. Yamada, A., S. C. Chung and K. Hinokuma, (2001), "Optimized Lifepo4 for Lithium Battery Cathodes," *J. Electrochem. Soc.*, 148, pp. A224-A229.

49. Yamada, Atsuo, Masao Yonemura, Yuki Takei, Noriyuki Sonoyama and Ryoji Kanno, (2005), "Fast Charging Lifepo4," *Electrochem. Solid-State Lett.*, 8, pp. A55-A58.

50. Yang, Shoufeng, Yanning Song, Peter Y. Zavalij and M. Stanley Whittingham, (2002), "Reactivity, Stability and Electrochemical Behavior of Lithium Iron Phosphate," *Electrochem. Commun.*, 4, pp. 239-244.

51. Yang, S., P. Y. Zavalij and M. Stanley Whittingham, (2001), "Hydrothermal Synthesis of Lithium Iron Phosphate Cathodes," *Electrochem. Commun.*, 3, pp. 505-508.

52. Yao, Masaru, Kazuki Okuno, Tsutomu Iwaki, Masahiro Kato, Shigeo Tanase, Katsuji Emura and Tetsuo Sakai, (2007), "Lifepo4-Based Electrode Using Micro-Porous Current Collector for High Power Lithium Ion Battery," *J. Power Sources*, 173, pp. 545-549.

53. Yu, D. Y. W., K. Donoue, T. Inoue, M. Fujimoto and S. Fujitani, (2006), "Effect of Electrode Parameters on Lifepo4 Cathodes," *J. Electrochem. Soc.*, 153, pp. A835-A839.

54. Iltchev, N., Y. K. Chen, S. Okada and J. Yamaki, (2003), "Lifepo4 Storage at Room and Elevated Temperatures," *Journal of Power Sources*, 119, pp. 749-754.

55. Thackeray, Michael, (2002), "Lithium-Ion Batteries: An Unexpected Conductor," *Nature Materials*, 1, pp. 81-82.

56. Doeff, M. M., Y. Q. Hu, F. McLarnon and R. Kostecki, (2003), "Effect of Surface Carbon Structure on the Electrochemical Performance of Lifepo4," *Electrochemical And Solid State Letters*, 6, pp. A207-A209.

57. Dominko, R., M. Bele, M. Gaberscek, M. Remskar, D. Hanzel, S. Pejovnik and J. Jamnik, (2005), "Impact of the Carbon Coating Thickness on the Electrochemical Performance of Lifepo4/C Composites," *J. Electrochem. Soc.*, 152, pp. A607-A610.

58. Franger, S., F. Le Cras, C. Bourbon and H. Rouault, (2002), "Lifepo4 Synthesis Routes for Enhanced Electrochemical Performance," *Electrochem. Solid-State Lett.*, 5, pp. A231-A233.

59. Hsu, K. F., S. Y. Tsay and B. J. Hwang, (2004), "Synthesis and Characterization of Nano-Sized Lifepo4 Cathode Materials Prepared by a Citric Acid-Based Sol-Gel Route," *J. Mater. Chem.*, 14, pp. 2690-2695.

60. Hu, Guo-rong, Xin-long Zhang, Zhong-dong Peng, Gang Liao and Xiao-yuan Yu, (2004), "Synthesis and Characterization of Lifepo4/C Composite Used as Lithium Storage Electrodes," *Transactions of Nonferrous Metals Society of China*, 14, pp. 237-

240.

61. Hu, Y. Q., M. M. Doeff, R. Kostecki and R. Finones, (2004), "Electrochemical Performance of Sol-Gel Synthesized Lifepo4 in Lithium Batteries," *J. Electrochem. Soc.*, 151, pp. A1279-A1285.

62. Huang, H., S. C. Yin and L. F. Nazar, (2001), "Approaching Theoretical Capacity of Lifepo4 at Room Temperature at High Rates," *Electrochem. Solid-State Lett.*, 4, pp. A170-A172.

63. Kwon, Sang Jun, Cheol Woo Kim, Woon Tae Jeong and Kyung Sub Lee, (2004), "Synthesis and Electrochemical Properties of Olivine Lifepo4 as a Cathode Material Prepared by Mechanical Alloying," *Journal of Power Sources*, 137, pp. 93-99.

64. Liao, X. Z., Z. F. Ma, L. Wang, X. M. Zhang, Y. Jiang and Y. S. He, (2004), "Novel Synthesis Route for Lifepo4/C Cathode Materials for Lithium-Ion Batteries," *Electrochemical And Solid State Letters*, 7, pp. A522-A525.

65. Myung, Seung-Taek, Shinichi Komaba, Norimitsu Hirosaki, Hitoshi Yashiro and Naoaki Kumagai, (2004), "Emulsion Drying Synthesis of Olivine Lifepo4/C Composite and Its Electrochemical Properties as Lithium Intercalation Material," *Electrochim. Acta*, 49, pp. 4213-4222.

66. Piana, M., B. L. Cushing, J. B. Goodenough and N. Penazzi, (2003), "New Sol-Gel Synthetic Route to Phospho-Olivines as Environmentally Friendly Cathodes for Li-Ion Cells," *Annali Di Chimica*, 93, pp. 985-995.

67. Prosini, P. P., D. Zane and M. Pasquali, (2001), "Improved Electrochemical Performance of a Lifepo4-Based Composite Cathode," *Electrochim. Acta*, 46, pp. 3517-3523.

68. Wang, C. W., A. M. Sastry, K. A. Striebel and K. Zaghib, (2005), "Extraction of Layerwise Conductivities in Carbon-Enhanced, Multilayered Lifepo4 Cathodes," *J. Electrochem. Soc.*, 152, pp. A1001-A1010.

69. Yang, Jingsi and Jun John Xu, (2005), "Novel Non-Aqueous Sol-Gel Synthesis of Carbon-Coated Limpo4 (M = Fe, Mn, Co) for Lithium Ion Batteries," *Mater. Res. Soc. Symp. Proc.*, 835, pp. 303-308.

70. Yang, J. S. and J. J. Xu, (2004), "Nonaqueous Sol-Gel Synthesis of High-Performance Lifepo4," *Electrochemical And Solid State Letters*, 7, pp. A515-A518.

71. Yang, J. S. and J. J. Xu, (2006), "Synthesis and Characterization of Carbon-Coated Lithium Transition Metal Phosphates Limpo4 (M = Fe, Mn, Co, Ni) Prepared Via a Nonaqueous Sol-Gel Route," *J. Electrochem. Soc.*, 153, pp. A716-A723.

72. Zane, Daniela, Maria Carewska, Silvera Scaccia, Francesco Cardellini and Pier Paolo Prosini, (2004), "Factor Affecting Rate Performance of Undoped Lifepo4," *Electrochim. Acta*, 49, pp. 4259-4271.

73. Sides, Charles R., Fausto Croce, Vaneica Y. Young, Charles R. Martin and Bruno Scrosati, (2005), "A High-Rate, Nanocomposite Lifepo4/Carbon Cathode," *Electrochem. Solid-State Lett.*, 8, pp. A484-A487.

74. Herle, P. S., B. Ellis, N. Coombs and L. F. Nazar, (2004), "Nano-Network Electronic Conduction in Iron and Nickel Olivine Phosphates," *Nature Materials*, 3, pp. 147-152.

75. Ni, J. F., H. H. Zhou, J. T. Chen and X. X. Zhang, (2005), "Lifepo4 Doped with Ions Prepared by Co-Precipitation Method," *Mater. Lett.*, 59, pp. 2361-2365.

76. Ni, J. F., H. H. Zhou, J. T. Chen and X. X. Zhang, (2005), "Improvement of Lifepo4 Electrochemical Performance by Doping Metal Oxides," *Chinese Journal Of Inorganic Chemistry*, 21, pp. 472-476.

77. Shi, Siqi, Lijun Liu, Chuying Ouyang, Ding-sheng Wang, Zhaoxiang Wang, Liquan Chen and Xuejie Huang, (2003), "Enhancement of Electronic Conductivity of Lifepo4 by Cr Doping and Its Identification by First-Principles Calculations," *Physical Review B: Condensed Matter and Materials Physics*, 68, pp. 195108/195101-195108/195105.

78. Wang, G. X., S. Bewlay, J. Yao, J. H. Ahn, S. X. Dou and H. K. Liu, (2004), "Characterization of Limxfe1-Xpo4 (M=Mg, Zr, Ti) Cathode Materials Prepared by the Sol-Gel Method," *Electrochemical And Solid State Letters*, 7, pp. A503-A506.

79. Wang, G. X., S. L. Bewlay, K. Konstantinov, H. K. Liu, S. X. Dou and J. H. Ahn, (2004), "Physical and Electrochemical Properties of Doped Lithium Iron Phosphate Electrodes," *Electrochim. Acta*, 50, pp. 443-447.

80. Yang, Mu-Rong, Wei-hsin Ke and She-huang Wu, (2007), "Improving Electrochemical Properties of Lithium Iron Phosphate by Addition of Vanadium," *J. Power Sources*, 165, pp. 646-650.

81. D'Epifanio, A., F. S. Fiory, S. Licoccia, E. Traversa, B. Scrosati and F. Croce, (2004), "Metallic-Lithium, Lifepo4-Based Polymer Battery Using Peo-Zro2 Nanocomposite Polymer Electrolyte," *J. Appl. Electrochem.*, 34, pp. 403-408.

82. Kang, Hsun Yueh, Sue Lein Wang, Ping Ping Tsai and Kwang Hwa Lii, (1993), "Hydrothermal Synthesis, Crystal Structure and Ionic Conductivity of Silver Vanadium Oxide Phosphate, Ag2vo2po4: A New Layered Phosphate of Vanadium(V)," *J. Chem. Soc., Dalton Trans.*, pp. 1525-1528.

83. PDF No. 01-081-2149, in *Calculated from FIZ#73580 3/18/08 by Jade.*

163e Stabilized Silicon Nanoparticles for High Capacity Li-Ion Battery Anode

Jeong-Kyu Lee, Michael N. Missaghi, Mayfair C. Kung, and Harold H. Kung. Chemical and Biological Engineering, Northwestern University, 2145 Sheridan Road E136, Evanston, IL 60208-3120

Increasing charge capacity and cycling stability of rechargeable Li ion batteries could greatly expand their potential applications, especially in transportation and as storage device for intermittent power sources. Compared to the graphite anodes currently in use, Si has much higher theoretical gravimetric and volumetric charge storage capacities ($Li_{4.4}Si \approx$ 4,200 mAh/g, 8,500 mAh/mL) and is an attractive candidate for high capacity lithium ion batteries. Because of the high surface to volume ratio, Si nanoparticles offer the potential of fast charge-discharge kinetics. However, sintering and fracturing of Si nanoparticles after cycling degrade the performance rapidly. We have explored stabilizing the Si nanoparticles by functionalizing the particle surface with phenolic hydrocarbon fragments which can be used in a simple solution phase in-situ sol-gel polymerization process to produce a homogeneous dispersion of Si in a Si-C composite after carbonization. The Si-C nanocomposite thus obtained is highly mesoporous (pore volume $\approx$ 0.7 cm^3/g, BET surface area $\approx$ 300~900 m^2/g), and initial results suggest that such composites are much more stable during cycling. The results of physical characterization, electrochemical behavior, and cycling stability of anodes prepared from Si nanoparticles, functionalized Si nanoparticles, and Si-C composite will be compared and reported. The advantages of such preparation will be discussed.

163f Developing Novel Microporous Carbons from Sugars – Morphology and Microstructure

J.E. St.Dennis[1], Pradeep Venkataraman[1], Bhanukiran Sunkara[1], **Vijay T. John**[1], Gary L. McPherson, Jibao He[2], Camille Y. Jones[3], Robert S. Currier[4], Steven J. Obrey[4], and Vladimir L. Kolesnichenko[5]. (1) Department of Chemical and Biomolecular Engineering, Tulane University, 300 Lindy Boggs Bldg., New Orleans, LA 70118, (2) Central Instrumental Facilities, Tulane University, New Orleans, LA 70118, (3) Department of Chemistry, Hamilton College, 198 College Hill Road, Clinton, NY 13323, (4) Chemistry Division, Los Alamos National Laboratory, MS J565, P.O. Boc 1663, Los Alamos, NM 87545, (5) Department of Chemistry, Xavier University of Louisiana, 1 Drexel Drive, New Orleans, LA 70125

Novel carbon materials, such as nano-tubes, fullerenes and uni-dimensional structures, and their applications to gas storage and as photo-voltaic devices have received significant attention in recent years. Different morphologies of carbons with varying degrees of crystallinity and microstructure can be obtained depending on the precursor solution used and the synthesis techniques employed. Carbons derived from sugars and carbohydrates as precursors have recently been of much interest. Hydrolysis of simple sugars and their derivatives followed by pyrolysis yields particles with spherical morphologies and graphitized nano-porous surfaces. Cyclodextrins, or cycloamylases, are cyclic oligosaccharides derived from the hydrolysis of amylase fractions of starch. They constitute a family of cyclic molecules with varying number of D-glucopyranonsyl residues, the abundant being α-, β- and γ- types containing six, seven and eight of these residues, respectively. The unique toroidal shape and distribution of hydrophilic and hydrophobic groups in these molecules make them attractive as host molecules for inclusion compounds as well as interesting precursors for carbon materials. While simple sugars have been studied extensively, to our knowledge, carbons derived from cyclodextrins have not yet been reported.We report the synthesis of nano-porous carbon microfibers that are on average 1 μm long and 20 nm wide. These novel carbons have surface areas of 400 – 500 m2/g and average pore size of 2-3 nm. While there is evidence of self-assembly of cyclic oligosaccharides at elevated temperatures, it has also been demonstrated that pyrolysis for extended time scales leads to formation of graphitic structures. These results provide an opportunity to better understand the self-assembly of cycloamylases and establish these materials as potential candidates for applications in gas storage as well as anodic materials for electrochemical reactions.

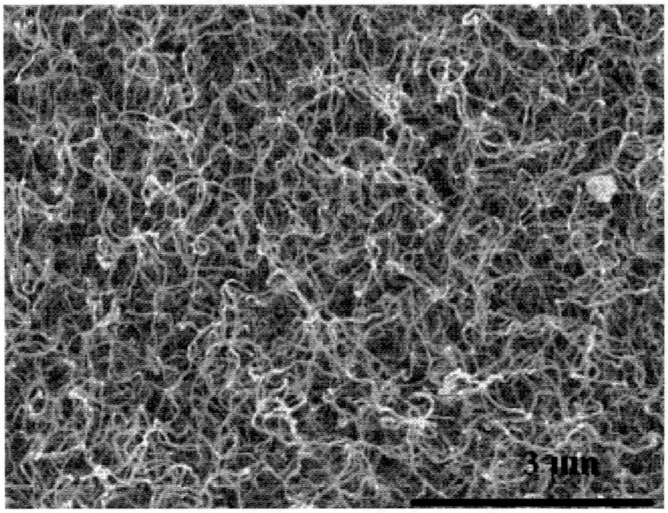
3 μm

608a Germanium Nanocrystals : Colloidal Synthesis and Optical Characterization

Doh C. Lee[1], Jeffrey M. Pietryga[2], Istvan Robel[2], Richard D. Schaller[2], and Victor I. Klimov[2]. (1) Los Alamos National Laboratory, MS-J567, Los Alamos, NM 87545, (2) C-PCS, Chemistry Division, Los Alamos National Laboratory, MS-J567, Los Alamos, NM 87545

Interest in fabricating solar cells comprised of semiconductor nanocrystals has grown in the past few years. Because of solubility of colloidal nanocrystals, the process involving deposition of active layers can be cheaper and more efficient than current thin-film technology. Colloidal inorganic nanocrystals offer an added advantage over organic polymers: nanocrystals passivated with capping ligands present better stability in air. Germanium is a particularly intriguing candidate for the use in photovoltaic cells, because of its low toxicity compared to II-VI semiconductors, the relative abundance, and its compatibility with current microelectronics technology. Moreover, the lifetime of excitons in germanium is longer than many other II-VI semiconductors suggested as the alternative to silicon technology. However, the use of germanium in solar cells has been limited due in part to the indirect nature of its band gap. Also, germanium is susceptible to oxidation, yielding electronically and chemically unstable germanium oxides. Nanocrystals with the size below exciton Bohr radius exhibit quantum confinement, which in the case of germanium could solve the indirect band gap problem. By growing nanocrystals with a stable surface coating, one could prevent the growth of oxide layers on the surface. Yet, the capability of synthesizing colloidal germanium nanocrystals is lacking, mainly due to the strong covalent bond of germanium and the need for high temperatures to break chemical bonds in precursors. The quest for facile syntheses that hints the growth of germanium nanocrystals with direct band-gap has thus intensified. We synthesized germanium nanocrystals via wet chemistry in a Schlenk setup. Germanium diiodide was used as the germanium precursor, and variables such as solvents, reducing agents, injection temperature, reaction temperature, reaction time and capping ligands were changed to monitor the best synthetic conditions. Capping ligands appear to affect the oxidation of germanium: surface passivation with covalently-bonded capping ligands allows the nanocrystals to be more resistive to even a severe oxidative environment. Colloidal germanium nanocrystals were characterized with transmission electron microscopy, x-ray diffraction, fluorometers operated at near-infrared or visible energy range, and ultrafast laser spectroscopy. We have synthesized the first germanium nanocrystals with strong infrared emission, with quantum yields as high as 10%.

608b Self-Assembly of Polythiophene Block Copolymers

Bryan Boudouris, Department of Chemical Engineering and Materials Science, University of Minnesota, Minneapolis, MN 55455, Marc A. Hillmyer, Department of Chemistry, University of Minnesota, 207 Pleasant St. SE, Minneapolis, MN 55455, and C. Daniel Frisbie, Chemical Engineering & Materials Science, University of Minnesota, 151 Amundson Hall, 421 Washington Ave S.E., Minneapolis, MN 55455.

Polythiophenes have attracted much attention in the field of organic electronics due to their high charge transport performance, their chemically tunable electronic properties, and their processibility from numerous solvents. In particular, poly(3-hexylthiophene) (P3HT) has become the most common electron-donating material in organic photovoltaics (OPVs) and recent advances in the fabrication and post-processing of P3HT–fullerene bulk heterojunction solar cells have allowed for devices with power conversion efficiencies of up to 5% to be consistently realized. This increase in efficiency is thought to arise from enhanced microphase separation of the active layer components and improved ordering in the structure of the film. These two factors allow for better separation of the bound electron-hole pair (exciton) that is generated by absorption of light in the active layer. An understanding of how exciton dissociation and the internal morphology of the active layer affects device performance would facilitate cell optimization and ultimately lead to higher efficiencies.

To more systematically study the effects of domain size and morphology on device performance, we have synthesized ABA-type triblock copolymers polylactide-b-poly(3-hexylthiophene)-b-polylactide (PLA-P3HT-PLA) by using a difunctional, alcohol-terminated poly(3-hexylthiophene) as a macroinitiator for the ring-opening polymerization of D,L-lactide and have thoroughly characterized the molecular and thermal properties of these novel materials. We have observed that when the P3HT moiety is amorphous the polymers self-assemble into ordered microstructures (domain spacing ~40 nm) both in the bulk (via SAXS) and in thin films (via AFM). We have previously shown that the polylactide domains can be selectively etched from the polythiophene matrix creating templates that could lead to all-organic ordered bulk heterojunction devices upon backfilling with an electron-accepting material.

608c Molecular Design for Organic Solar Cells

Liping Huang, Department of Chemical and Biomolecular Engineering and Department of Physics, North Carolina State University, Raleigh, NC 27606, Dario Rocca, Department of Chemistry, University of California, Davis, Davis, CA 95616, Stefano Baroni, SISSA and CNR-INFM DEMOCRITOS National Simulation Center, Trieste, Italy, Keith E. Gubbins, Chemical and Biomolecular Engineering and Center for High Performance Simulation (CHiPS), North Carolina State University, Raleigh, NC 27695-7905, and Marco Buongiorno Nardelli, Department of Physics, North Carolina State University, Raleigh, NC 27695.

Organic semiconductors are promising candidates for making low-cost solar cells because of their easiness to process. However, they usually have limited response to the solar spectrum and very low charge carriers mobility. We carried out first-principles calculations at the density functional theory (DFT) and time-dependent density functional theory (TDDFT) levels to understand the effect of chain length, functional groups and molecular packing on the optical absorption spectrum and charge carrier motility of acenes. Our calculations demonstrate that these properties not only depend on the molecular identity but also on the molecular packing. By designing the interaction between metal substrates and the first layer of acene molecules, they can be packed in a cofacial (face to face) fashion instead of the conventional herringbone (face to edge) arrangement. Acenes packed in the cofacial way show improved response to the solar spectrum and high charge carriers mobility. This could open the door to a new route of materials design for cost-effective organic solar cells.

608d Perylene-Thiophene Based Donor-Acceptor Oligomers for Molecular Scale Photovoltaics

Jibin Sun, Chemistry and Chemical Engineering Department, University of California, Berkeley, Tan Hall B81, Chemistry and Chemical Engineering Department, UC, Berkeley, Berkeley, CA 94720-1460, T. Don Tilley, Department of Chemistry, University of California, Berkeley and Lawrence Berkeley National Laboratories, Berkeley, CA 94720, and Rachel A. Segalman, University of California at Berkeley, Dept of Chemical Engineering, 201 D Gilman, Berkeley, CA 94720-1462.

Organic photovoltaic solar cells represent a cheap and lightweight alternative energy source that hold promise for solving the world's energy needs. The highest efficiency devices to date are heterojunction solar cells, in which donor (D) and acceptor (A) semiconductors are blended together. The efficiencies of the devices are strongly dependent on the properties of the interfaces on a scale of 5-10 nm, where excitons are created. Control of this length scale and the interface morphology has proven to be quite difficult. D-A oligomers provide an elegant route to achieve this goal by variation of several tunable parameters, such as the length of oligomer blocks, the functionality of each block, and the bridging moieties. We have prepared a series of perylene-thiophene-based oligomers to study the relationships between molecular structure, thin film morphology, and device performance.

We started with one perylene molecule linked to a thiophene dimer or trimer by a double bond. This p-n junction molecule shows high thermal stability and can be transferred to different substrates by gas phase evaporation methods. With different R groups at one terminal end of the perylene, we observed disordered, short range ordered, and highly ordered structures on an Au (111) surface by STM studies. Grazing-angle incidence X-ray diffraction (GIXD) has also been applied to understand the three dimensional structures of the films. This structural information provides significant understanding of the thin film morphology, and should help in the optimization of device performance.

In addition, a series of p-n junction oligomers have been prepared. In this type of molecular assembly, perylene-naphthalene (A) is connected by different bridging units to hexathiophene (D). These molecules are designed to help answer some fundamental questions about the structural requirements of charge separations and recombination: What is the minimum length to achieve molecular dipolar character? What is the function of various bridges? Does the configuration of the bridge affect the morphology of the thin film? The synthesis of these molecules will be discussed, along with the electrical and optical properties of the new materials, in both solution and the solid state.

608e Mobility Measurement In Organic Charge Transport Materials by Charge-Retraction Time-of-Flight

Jason U. Wallace[1], Ralph H. Young[2], Ching W. Tang[1], and Shaw H. Chen[1]. (1) Chemical Engineering, University of Rochester, 240 East River Road, Rochester, NY 14623, (2) Eastman Kodak Company, 1999 Lake Avenue, Rochester, NY 14620-2110

In both organic light-emitting diodes (OLEDs) and organic photovoltaics (OPVs), the mobility of charge carriers in the various organic layers is a key parameter. A number of methods have been developed to measure this mobility, each with their own advantages and disadvantages. In the transient space-charge-limited current (SCLC) method, the sample is subjected to a rectangular voltage pulse, and the resulting current exhibits a characteristic peak “cusp” at a space-charge perturbed time-of-flight (TOF) before settling down to a steady-state level. Both the TOF and the steady-state level can be used to determine the bulk mobility. The SCLC technique, however, requires that the injecting contact be ohmic, a requirement that is rarely met. Transient electroluminescence is another technique, and one that lessens the demand on these injecting contacts somewhat. In transient electroluminescence measurements, a voltage pulse is applied to a suitably structured OLED, and the onset of light emission is evaluated for the TOF; however, analysis is complicated by the kinetics of supplying carriers of the opposite sign. In these three methods, it is difficult to reliably extract any information beyond the effective mobility of the faster charge carriers. Charge extraction in a linearly increasing voltage is principally applicable to samples with appreciable conductivity in the dark, but has begun to be combined with photoexcitation. While all these methods remain in use, the most popular technique for studying charge carrier mobility in organic materials is photocurrent time-of-flight. Here a thin layer of a sample is photo-excited by a flash of light, often a pulsed laser, and the resultant charges are swept across the sample by an applied electric field. Due to the penetration depth of the light, charges are generated well within the sample layer, often necessitating samples of several microns in thickness. Fabricating such thick samples can be materials intensive, and it is challenging to keep the thickness and morphology uniform. In addition, such thick films are potentially unrepresentative of the layers used in working devices, which are typically one-tenth to one-hundredth as thick.

In this presentation, a relatively new method for mobility measurement will be discussed. Charge-retraction time-of-flight (CR-TOF) involves the injection, blocking, accumulation, and retraction of charge carriers to realize a transient functionally equivalent to those obtained by a traditional photocurrent TOF experiment, giving information about both charge carriers. It has been validated by measurement of the hole mobility of two well-known compounds, 4,4',4"-tris-[N-(3-methylphenyl)-N-phenylamino]triphenylamine and 4,4'-bis[N-(1-naphthyl)-N-phenylamino]biphenyl, as well as preliminary measurement of the electron mobility of 4,7-biphenyl-phenanthroline, with appropriate blocking layers. With this technique, a sample layer thickness of less than 300 nm can be used because of the thinness of the charge sheet accumulated at the blocking layer present. The availability of this blocking layer is the primary constraint for the method; an excellent one has been found for blocking holes, while a feasible one has been found for electrons. Two additional points of interest have emerged from developing this method. First, the arbitrariness associated with evaluating the transition voltage at organic-organic interfaces was addressed by examining the excess charges collected as a function of charging voltage. Second, by changing the charging time, charges could be retracted while in transit to the blocking layer, thus providing opportunities to study their distribution and possibly trapping or leakage at the blocking interface.

608f Processing of Transparent Conducting Films from Colloidal Suspensions of Carbon Nanotubes

Mainak Majumder[1], Clint Rendall[1], Budhadipta Dan[2], Howard K. Schmidt[3], and Matteo Pasquali[4]. (1) Chemical and Biomolecular Engineering, Rice University, 6100 Main Street, MS-362, Houston, TX 77005, (2) Physics and Astronomy, Rice University, 6100 Main Street, MS-362, Houston, TX 77005, (3) Chemical and Biomolecular Engineering, Rice University, 6100 Main Street, Houston, TX 77005, (4) Chemical and Biomolecular Engineering, Chemistry, Rice University, 6100 Main Street, MS-362, Houston, TX 77005

CNT-films are thought as possible replacement to Indium Tin Oxide for a variety of uses, including flat-panel displays, lasers, and solar cells. We report the fabrication of optically transparent and electrically conducting thin films of single-walled carbon nanotubes (SWNT) from surfactant stabilized CNT suspensions using two different processes: Spray-Coating and Roll-Coating - two scalable coating techniques used industrially.

Focus of this talk will be on the spray-coating method. The advantage of the spray-coating process is its applicability to irregular surfaces. In this method, an ultra-sonic droplet generator was utilized to form uniform droplets of the SWNT suspensions, which was deposited on a glass-substrate kept at a temperature of 150-200 °C. Optimization of several processing parameters such as choice of surfactant and solvent, sonication time for dispersion synthesis, drying conditions, surfactant washing are critical for the fabrication of these films. The films were also treated with vapors of strong acids which lead to significant improvement in properties. Films with sheet resistance of 150 Ohm/sq at a transparency of 50% were obtained. Processing-property correlation of spray-coated CNT films will be discussed.

704a Nanoscale Heterojunction Engineering to Grow High-Quality Ge on Si for Multijunction Solar Cells

Darin Leonhardt[1], Josephine Sheng[1], Thomas Vandervelde[2], Jeffrey Cederberg[3], Malcolm Carroll[3], and **Sang M. Han**[1]. (1) University of New Mexico, 1 University of New Mexico, MSC01 1120, Albuquerque, NM 87131-0001, (2) Electrical Engineering & Computer Engineering, Center for High Technology Materials, 1313 Goddard SE, Albuquerque, NM 87106, (3) Sandia National Laboratories, Sandia National Laboratories, P.O. Box 5800, Albuquerque, NM 87185-0601

In an effort to reduce the manufacturing cost of multijunction solar cells, we have scaled up a process to grow low-defect-density Ge films on 2-inch-diameter Si substrates. This growth technique makes use of nanoscale heterojunction engineering to minimize the interfacial strain density. The engineered substrates may potentially replace the Ge wafers that are currently used in multijunction solar cell fabrication, if the Ge film's bulk and surface quality can match that of the epi-ready Ge wafers. We will present results for the scaled-up process of Ge film production, including key aspects of the nucleation process and film characterization, using transmission electron microscopy and etch pit counting. Next, we present our efforts to produce a high-quality surface finish, using chemical-mechanical planarization, and method for cleaning and passivating the Ge surface. Additionally, results of GaAs film growth on our engineered substrates will be presented and compared to growth on Ge and GaAs wafers, both offcut and nominal. We find that the offcut wafers effectively eliminate anti-phase domains in the GaAs. We also observe room-temperature photoluminescence from the GaAs epilayer grown on our engineered Ge/Si substrates. Lastly, future work and directions will be discussed in light of our findings.

704b Plasmonics-Based Design of Nanostructured Materials for Solar Energy Harvesting

J. Trice[1], R. Kalyanaraman[1], H. Garcia[2], and **R. Sureshkumar**[3]. (1) Department of Physics, Washington University, Saint Louis, MO 63130, (2) Department of Physics, Southern Illinois University, Edwardsville, IL 62026, (3) Department of Energy, Environmental and Chemical Engineering and the Center for Materials Innovation, Washington University, Saint Louis, MO 63130

Metallic nanoparticles embedded in dielectrics permit enhanced capture of light at specific wavelengths through excitation of plasmons, i.e. the quanta of coherent and collective oscillations of large concentrations of nearly free electrons. In order to maximize the potential of such enhanced absorption in useful tasks, such as the generation of carriers in photocatalysis and semiconductors, it is important to be able to predict and design plasmonic nanocomposites with desired wavelength-dependent optical response. Recently, a mixing approach formulated by Garcia and coworkers [Phys. Rev. B, 75, 045439 (2007)] has been successfully applied to model the experimentally measured broadband optical absorption for ternary nanocomposites containing alloys or mixtures of two metals (from Ag, Au or Cu) in SiO2 dielectric. In this work we present the broadband optical behavior of important optical coating dielectrics Si3N4 and SiO2 and photocatalyst TiO2 containing various configuration of nanoparticles of Al, Au, Ag, or Cu. The spectral behavior of various combinations of the metallic species in the dielectrics was optimized to show its use as an anti-reflecting coating on Si or strong multiple plasmonic absorption peaks. The applications of such nanocomposite materials in solar energy harvesting and spectral sensing are also presented and discussed.

704c Band-Edge Engineered Hybrid SnO2 Nanowires Electrodes for Dye Sensitized Solar Cells

Suresh Gubbala[1], **Vidhya Chakrapani**[2], Vivekanand Kumar[1], Heather Rypkema[3], and Mahendra K. Sunkara[1]. (1) Department of Chemical Engineering, University of Louisville, 106 Ernst Hall, Department of Chemical Engineering,, University of Louisville, Louisville, KY 40292, (2) Chemical Engineering, University of Lousiville, Room 311, Ernst Hall, Univeristy of Louisville, Louisville, KY 40292, (3) Chemistry, Univeristy of Louisville, Louisville, 40292

Dye sensitized solar cells (DSSC's) based on wide band gap semiconducting oxide nanomaterials have gained increased importance in the last few years due to their high solar to electric conversion efficiencies at low costs. Of the different types of nanostructures, highly crystalline nanowires offer unique advantages over nanoparticle networks by providing a direct and fast pathway for electron transport. In this work, we show that SnO_2 nanowire based dye sensitized solar cells exhibit an open circuit voltage of 560 mV, which is 200 mV higher than that using SnO_2 nanoparticle based cells. The nanowires were employed in hybrid structures consisting of highly interconnected SnO_2 nanowire matrix coated with TiO_2 nanoparticles, which showed an open circuit voltage of 720 mV and an efficiency of 4.1% compared to 2.5% obtained with pure SnO_2 nanowire matrix. Transient photovoltage and photocurrent measurements of hybrid electrodes show that the recombination time constant is ~100 times higher and transport time constants is ~10 times lower than that of TiO_2 nanoparticles. The higher efficiency observed for DSSC's based on hybrid structure is attributed to the band edge positions of SnO_2 relative to that of TiO_2 and faster electron transport in SnO_2 nanowires.

704d Erbia-Coated Quartz Fibers as a Selective Emitter for Low Temperature Tpv Applications

Mohannad T. Aljarrah and Edward A. Evans. Chemical and Biomolecular Engineering, University of Akron, 302 Buchtel common, Akron, OH 44325

Thermophotovoltaic generators convert radiative thermal energy to electrical energy. Basically, any source of thermal energy could be used to heat the emitter. One of the major challenges for photovoltaic generators is the mismatch between the emitter and the PV cell bandgap. Rare-earth oxides are known to radiate infrared radiation in a very narrow range that can be matched with different types of PV cells. In this study, selective emitters made by erbia-coated quartz fibers using dip coating technique were made. The fabricated emitter utilizes the high thermal and mechanical stability, high surface area to volume ratio of the fibers and the selective radiation of the erbia coating. A uniform coating layer with strong adherence to the fibers was achieved by controlling the concentration of erbium oxide and the viscosity of the dipping solution. The spectral emissivity was measured using a specially designed FTIR setup

704e Self-Assembled Biomimetic Antireflection Coatings

Chih-Hung Sun[1], Wei-Lun Min[1], Nicholas Linn[1], Bin Jiang[2], and **Peng Jiang**[1]. (1) Chemical Engineering, University of Florida, Gainesville, FL 32611-6005, (2) Mathematics and Statistics, Portland State University, Portland, OR 97201

Antireflection coatings (ARCs) are widely utilized in reducing the unwanted reflection from flat-panel displays, car dashboards, optical components, and solar cells. Inspired by the microstructured cornea of some nocturnal moths, subwavelength-structured ARCs have been extensively explored. However, current lithography-based fabrication techniques in creating subwavelength features are costly and are limited by either low resolution or small sample size. Here we report several cheap and scalable bottom-up techniques for fabricating large-area moth-eye ARCs on transparent (e.g., glass and polymer) and semiconductor (e.g., Si, GaAs, and GaSb) substrates. All these techniques are based on a simple spin-coating technique that enables the creation of wafer-scale colloidal crystals with non-close-packed structures. The resulting ARCs exhibit superior broadband antireflection properties that are useful in developing high-efficiency solar cells and detectors. Two numerical methods – thin-film multilayer and rigorous coupled-wave analysis models have also been developed to simulate the specular reflection from the templated subwavelength structures.

Web Page: www.che.ufl.edu/jiang

704f Biomimetic Solar Energy Conversion by Photosystem I Films

Christopher J. Faulkner, Peter N. Ciesielski, and G. Kane Jennings. Department of Chemical and Biomolecular Engineering, Vanderbilt University, Box 1604 Station B, Nashville, TN 37235

Photosystem I (PSI) has drawn widespread interest for use in biomimetically inspired energy conversion devices upon extracting it from plants or cyanobacteria and assembling it at electrode surfaces. We report that ~1 micron thick films of PSI on gold electrodes provide an order-of-magnitude improvement in photo-generated current over that of a PSI monolayer although only one layer of PSI is in contact with the electrode in each case. We have investigated the mechanism for this dramatic improvement by monitoring the effect of PSI film thickness and electrochemical mediator on photoinduced current. The partitioning of hydrophobic mediators into the moderately hydrophobic PSI film creates an elevated concentration of electron acceptors to rapidly draw current from the interfacial, bound PSI layer as well as electron donors to turn over the P700 reaction centers dispersed throughout the film. We will also discuss design considerations in the fabrication of a photoelectrochemical cell that contains these thin films of PSI. At present, these prototype cells can be used to power simple devices such as calculators.

AUTHOR INDEX

AUTHOR INDEX